# The Moral Molecule

www.**transworldbooks**.co.uk

*Also by Paul J. Zak*

The Moral Markets: The Critical Role of Values in the Economy

For more information on Paul J. Zak and his books,
see his website at www.moralmolecule.com

# The Moral Molecule

THE SOURCE OF LOVE AND PROSPERITY

## Paul J. Zak

BANTAM PRESS

LONDON · TORONTO · SYDNEY · AUCKLAND · JOHANNESBURG

TRANSWORLD PUBLISHERS
61–63 Uxbridge Road, London W5 5SA
A Random House Group Company
www.transworldbooks.co.uk

First published in Great Britain
in 2012 by Bantam Press
an imprint of Transworld Publishers

A CIP catalogue record for this book
is available from the British Library.

ISBN 9780593067499 (cased)
9780593067505 (tpb)

Addresses for Random House Group Ltd companies outside the UK
can be found at: www.randomhouse.co.uk
The Random House Group Ltd Reg. No. 954009

The Random House Group Limited supports the Forest Stewardship Council (FSC®), the
leading international forest-certification organization. Our books carrying the FSC label are
printed on FSC®-certified paper. FSC is the only forest-certification scheme endorsed by
the leading environmental organizations, including Greenpeace. Our paper procurement
policy can be found at www.randomhouse.co.uk/environment

Printed and bound in Great Britain by
CPI Group (UK) Ltd, Croydon, CR0 4YY

Designed by Daniel Lagin

2 4 6 8 10 9 7 5 3 1

MIX
Paper from
responsible sources
FSC® C016897

*To my daughters Alexandra and Elke who, through their love,*
*have made me a better and happier person*

# Contents

# CONTENTS

# Introduction

*Vampire Wedding*

It was a lovely day for a wedding, the English sun peeking from behind English clouds as the guests gathered in their finest. The ceremony was going to be at Huntsham Court, a Victorian manor house out in Devon, and it was set to begin in ten minutes. I was supposed to have shown up an hour ago.

I parked my rented Vauxhall in the gravel courtyard, left the engine running, jumped out in my lab coat for an immediate re-connoiter, then commandeered a guest to help me carry in the 150-pound centrifuge and 30 kilos of dry ice I'd brought with me in the car. With a second trip I carried in the syringes, 156 pre-labeled test tubes, tourniquets, alcohol preps, and Band-aids that I'd shipped from California.

The plan I'd worked out with Linda Geddes, the bride, was to take two samples—one blood draw immediately before the vows, and one immediately after—from a cross-section of friends and family in attendance. Within the wedding party itself, Linda's father was the only holdout; the mother of the groom had been ill, so we gave her a pass.

Now, taking blood at weddings is not a long-standing tradition in this part of England, or anywhere else that I know of. In this case, the bride was a writer for the *New Scientist* who'd been following my research. She was also known for throwing herself into her stories gonzo-style. One day out of the blue she invited me to fly across the Atlantic to see her get married, but it wasn't because we'd become such close pals. She wanted me to run an experiment to illustrate a point. Just for fun, she wanted to see if the emotional uplift of her wedding would alter the guests' levels of oxytocin (not to be confused with the often-abused painkiller OxyContin), the chemical messenger I'd been studying for the past several years. Oxytocin is known primarily as a female reproductive hormone, and usually it's associated less with wedding vows and champagne than with what, in an earlier time, often happened nine months after. Oxytocin controls contractions during labor, which is where many women encounter it as Pitocin, the commercially available synthetic version doctors inject in expectant mothers to induce delivery. Oxytocin is also responsible for the calm, focused attention mothers relish on their babies while breast-feeding. Then again, oxytocin is well represented—we hope—on wedding nights, because it helps create the warm glow both women and men feel during sex, or a massage, or even a hug.

Linda hadn't reached out to me because of anything new I had to say about oxytocin as the "birth hormone" or the "cuddle hormone," but because of an entirely different role I'd discovered for it. My research had demonstrated that this chemical messenger both in the brain and in the blood is, in fact, the key to moral behavior. Not just in our intimate relationships, but also in our business dealings, in politics, in society at large.

Which is a point, I realize, that might take some getting used to.

Am I actually saying that a single molecule—and, by the way, a chemical substance that scientists like me can manipulate in the lab—accounts for why some people give freely of themselves and others are coldhearted bastards, why some people cheat and steal and others you can trust with your life, why some husbands are more faithful than others, and, by the way, why women tend to be more generous—and nicer—than men?

In a word, yes.

Beginning in 2001, my colleagues and I conducted a number of experiments showing that when someone's level of oxytocin goes up, he or she responds more generously and caringly, even with complete strangers. As a benchmark for measuring behavior, we relied on the willingness of the people being tested to share real money in real time. To measure the increase in oxytocin, we took their blood and analyzed it.

Money, as everybody knows, comes in conveniently measurable units—nickels and dimes, tens and twenties—which meant that we were able to quantify the increase in generosity by the amount someone was willing to share. We were then able to correlate these numbers with the increase in oxytocin found in the blood. Later, to be absolutely certain that what we were seeing was not just an association but true cause and effect, we infused synthetic oxytocin into our study subjects' nasal passages—the next best thing to shooting it directly into their brains. As for cause and effect, we found that we could turn the behavioral response on and off like a garden hose.

But what our work demonstrated first and foremost is that you don't need to shoot a chemical up someone's nose, or have sex with them, or even give them a hug in order to create the surge in oxytocin that leads to more generous behavior. Fortunately, all you have to do to trigger this Moral Molecule is give someone a sign of trust. When one person extends himself to another in a trusting way, the person being trusted experiences a surge in oxytocin that makes her less likely to hold back, and less likely to cheat. Which is another way of saying that the feeling of being trusted makes a person more . . . trustworthy. Which, over time, makes other people more inclined to trust, which in turn . . .

If you detect the makings of an endless loop here that can feed back onto itself, creating what might be called a virtuous cycle—and ultimately a virtuous society—you're getting the idea. And that's what's so incredibly exciting about this research.

Obviously there's more to it, because no one chemical in the body functions all alone, and other factors from a person's life experience play a role as well. But as we'll see in the chapters ahead, oxytocin orchestrates the kind of generous and caring behavior that every culture, everywhere in the world, endorses as the right way to live, the cooperative, benign, pro-social way of living that every culture everywhere on the planet describes as "moral."

Which is not to say that oxytocin always makes us good, or always generous and trusting. In a rough-and-tumble world, unwavering openness and loving kindness would be like going around with a KICK ME sign on your back. Instead, the Moral Molecule works like a gyroscope, helping us maintain our balance between behavior based on trust, and behavior based on wariness and distrust. In this way oxytocin helps us navigate between the social benefits of openness—which are considerable—and the reasonable caution we need to avoid being taken for a ride.

It was oxytocin's ability to recognize and respond to the *precise* nature of human bonds and interactions that intrigued Linda the bride, so much so that she invited me to her wedding. She wanted to see how witnessing all the promises to be faithful, and caring, and committed would play out not in her guests' behavior, but in their blood.

Huntsham Court is about four hours west of London, hidden among little villages with names like Lower Washfield, Stoodleigh, and Clayhanger. There's a crumbling Anglo-Saxon church on the grounds, but the official part of the ceremony was going to take place in the manor house itself, an old hunting lodge saturated with the smell of wood fires and oak paneling and the mounted heads of long-dead animals.

After all my running back and forth like the cliché mad scientist, I settled into the space just off the main room that had been set aside for my portable blood lab—the centrifuge borrowed from the University of Exeter, the dry ice sent from London. To point the way for Helen, a nurse and friend of the bride's who'd agreed to draw the blood, someone hung a makeshift sign on the door that said SCIENCE LAB.

I was delighted to have a locally and legally qualified assistant, but when Helen showed up, she was in high heels and a beige silk dress, not the surgical scrubs or lab coat I guess I'd imagined. *No room for error here,* I thought.

We went over the protocol for the experiment, and I made sure all the equipment was turned on and ready. Then with my well-dressed colleague in tow, I went to find my first victim.

Luckily for me, Linda was running late herself. I found her in the bride's suite upstairs being fluffed and pampered by her mother and her bridesmaids, three young women dressed, appropriately enough for a "vampire" wedding, in bright crimson.

Linda and I had never actually met, but on this happy occasion she greeted me with hugs and kisses all the same.

I said, "You ready for this?"

She grinned nervously as her friend got down to business, placing the tourniquet on her arm and swabbing the skin.

"Not too keen on needles, actually," she said.

"Now you tell me," I replied, and reached for the smelling salts I had tucked away in my pocket just in case.

Through it all, neither bride, nor guest, nor yours truly fainted (truth be told, I love the sight of blood), and Linda's dedication to a good story didn't spoil her big day. Near as I could tell, the assembled friends and family seemed to think all this blood drawing business was actually quite a lark.

After the vows and the registry-signing inside the mansion, everyone went outside for the handfasting, a Celtic tradition involving personally crafted vows under a tree festooned (that's what they do in England) with colored ribbons, overseen, in this case, by a fellow journalist who happened to be Hindu. Covering all the bases, I suppose.

Then the wedding party came back in for more blood draws—twenty-four samples in just under ten minutes—and we were done. Linda and Nic, her good-natured groom, could move on to champagne, and dinner, and dancing on the lawn to the music of a swing band. But ever the science nerd, I was stuck back in the big house, spinning down the samples in the centrifuge, separating the serum and the plasma from the red blood cells, and fast-freezing the blood products I needed to analyze for changes in oxytocin. Then, with my test tubes nestled in their cushion of dry ice, I slipped out quietly and began the long trek back to London, and from there the even longer trek back to my lab at Claremont Graduate University in Southern California. It took two weeks (and about $500) for the samples to arrive via FedEx, and then another $2,000 for us to analyze the blood. But after all was said and done, the results showed just what we were hoping for, which was a simple snapshot of oxytocin's ability to read and reflect the nuances of a social situation, and thus become the monitor and key regulator of our moral behavior.

Everybody knows that marriage ceremonies are emotionally charged. That's why people cry at weddings. That's why the bad boys in *Wedding Crashers* showed up at so many of them—to pick up girls primed and ready to be warm and cuddly. But the blood samples at Huntsham Court showed us something much more interesting. The changes in individual oxytocin levels at Linda's ceremony could be mapped out like the solar system, with the bride as the sun. Between the first and second blood draws, which were only an hour apart, Linda's own level shot up by 28 percent. And for each of the other people tested, the increase in oxytocin was in direct proportion to the likely intensity of emotional engagement in the event. The mother of the bride? Up 24 percent. The father of the groom? Up 19 percent. The groom himself? Up 13 percent . . . and on down the line with siblings and friends with more peripheral roles to play.

But why, you may ask, would the groom's increase be less than his father's? We'll get into this sort of thing more deeply along the way, but testosterone is one of several other hormones that can interfere with the release of oxytocin. Not too surprising when you think about it, I also found that the groom's testosterone had surged 100 percent!

Our little study at the wedding had demonstrated, on the hoof, just the kind of graded and contingent sensitivity that allows oxytocin to guide us between trust and wariness, generosity and self-protection, not only in response to the official nature of relationships—my mother, my son-in-law, my dreaded classmate, a complete stranger—but in response to social cues in the moment. Should I feel safe and warm and cuddly in this crowd, or do I have to be on guard? Is this a situation in which oxytocin can call the tune, or is this an exchange in which survival will be better served by a surge of a stress hormone that will keep me on my guard? Or maybe it's a situation in which the very best outcome will result when oxytocin dominates in one party and there's a healthy dose of testosterone driving the other.

It's the sensitivity of oxytocin in its interaction with a range of other chemical messengers that helps to account for why human behavior is so infinitely complex—and why the bliss of the wedding day (and night) is often hard to maintain. (There's the old joke about the guy from Finland who couldn't understand why his wife was unhappy. "I told you that I loved you when I asked you to marry me," he said. "I don't see why I need to tell you again.")

But here's the much larger payoff from the much larger body of research my lab has conducted: After centuries of speculation about human nature, human behavior, and how we decide what is the right thing to do, here at last we have some news we can use— solid empirical evidence that illuminates the mechanism at the heart of the moral guidance system. As any engineer will tell you, understanding the basic mechanism is the first step toward improving a system's output. Which, when the output is moral behavior, is no trivial matter.

In just the past few years, new insights as to why people behave as they do have been flooding in from fields like behavioral economics, social neuroscience, neurotheology, evolutionary studies of altruism and cooperation, even happiness research. All these data suggest that, as a species, we are far less self-interested—and, on balance, generally far kinder and more cooperative—than the prevailing wisdom has ever acknowledged.

But up until now, this scientifically enhanced insight into human nature—the good as well as the bad—still begged the question: Given that humans can be both rational and irrational, ruthlessly depraved and immensely kind, shamefully self-interested as well as completely selfless, what *specifically* determines which aspect of our nature will be expressed when? When do we trust and when do we remain wary? When do we give of ourselves and when do we hold back? The answer lies in the release of oxytocin.

Oxytocin surges when people are shown a sign of trust, and/or when something engages what was once called "our sympathies," which is what we now call empathy. (I'll delve into how empathy does its thing in chapter 4.) When oxytocin surges, people behave in ways that are kinder, more generous, more cooperative, and more caring. But when scientists call these behaviors pro-social, that's actually just a geek-speak way of saying that they follow the Golden Rule: "Do unto others as you would have them do unto you." This book is going to show you why this oxytocin effect happens, when it happens, and how we can make it happen more often.

The fact that the Moral Molecule pries open the black box of human nature doesn't mean that there's nothing left for the philosophers and theologians to wrestle over. It's just that any discussion of free will or virtue seems a little pointless if it doesn't take into account everything science can contribute. And we've learned quite a lot since those early prophets tried to divine what God wanted us to do, and the philosophers tried to figure it out through the power of reason.

After all the theological debates and all the philosophical discourse and all the new evidence, one thing we know for sure is that humans are intensely social creatures. The human brain reacts more intensely to a human face than to any other object in the universe. That's because survival during our first years of life is entirely dependent on the goodwill of others—namely our parents—and their willingness to invest resources in us. Even after we're old enough to provide for ourselves, we continue to depend on a web of social cooperation to stay alive and healthy. We are, in fact, what zoologists call an obligatorily gregarious species, meaning that we thrive in groups, and that we don't do well, either physically or emotionally, for long periods alone. All of which helps to explain why we are so intensely interested not only in other people's facial expressions and emotions, but also in their behavior—who's doing what to whom, who's a stand-up guy, and who's a sleazeball hiding behind a fake smile. Oxytocin primes us to react appropriately, even when we have no idea that it's on the job.

In this book I'm going to explore oxytocin's influence on the individual, its influence on close personal relationships, and then its influence on society as a whole. Along the way we're going to see how various life experiences and different ways of thinking can alter the oxytocin effect. We'll also look into the influence of religion—a biggie when it comes to discussions of morality—as well as the influence of a market economy. In turn, we'll also discuss oxytocin's influence on those well-established institutions.

A consistent theme is going to be that, unless the release of oxytocin is impaired, the Golden Rule is a lesson the body already knows, and when we get it right we feel the rewards immediately. These range from better health, to a happier life, to—believe it or not—a more prosperous economy. And the vast majority of people don't have to be beaten over the head, don't have to listen to long sermons, and don't have to be threatened with hellfire and damnation to want to treat others well. To elicit that naturally occurring, benign behavior, all we have to do is to create the circumstances in which oxytocin can exercise its influence, which means, in large part, keeping other hormonal influences out of the way. Easier said than done, of course, but I think you'll agree that knowing how the system actually works is an excellent start.

We began the oxytocin story at a wedding, which is all the more appropriate because oxytocin, as you may recall, is a reproductive hormone. A biological link between sex and morality? What a concept.

Hundreds of millions of years ago, when sex first evolved, depending on the kindness of strangers was a good way to become lunch. "Big fish eat little fish" was the order of the day, every day. So how were two creatures supposed to get together to mate? They needed a chemical messenger that would make it safe to trust, by prompting benign behavior in response to trust. Sound familiar?

The role of trust is woven through everything we'll be discussing here. It even permeates the backstory of how I came to be doing this work. As I'll explain in greater detail later, I actually began my academic career building economic models of what makes countries prosper. My early work demonstrated that the most important factor in determining whether or not a society does well or remains impoverished is not natural resources, education, quality health care, or even the work ethic of its people. What matters most in determining economic outcomes is actually trustworthiness—a moral consideration. That's the insight that led me to the Moral Molecule.

Well before that, however, I'd become fascinated by the power of trust, mostly because I'd discovered the dangers of trusting too much. This happened when I was still a pretty naïve kid and became the victim of a classic con called the Pigeon Drop. That's where you might say my research career began. This book is where it's taken me so far.

# CHAPTER 1

# The Trust Game

*From Short Cons to the Wealth of Nations*

The scene of the crime was an ARCO station, in a sketchy neighborhood on the outskirts of Santa Barbara, where I had an after-school job pumping gas.

One day I was standing in the doorway to the office, feeling the breeze and waiting for the next customer to pull up to the pump, when a well-dressed but slightly worried-looking guy walked around from the side of the building.

"Maybe you can help me," he said. "I've got a job interview up in Goleta and I don't know what to do."

"What's up?" I said.

"Well, look . . ." He held out a small gift box from a fancy jewelry store in town. Then he opened the box, and inside was a pearl necklace, glimmering in the California sun.

"I just used your men's room and I found this on the floor. Amazing, huh? Has anybody called?"

"Not yet."

"Man, that's a nice piece of jewelry. Somebody's really going to be upset they lost this. What do you think we should do? I can't just keep it."

We both stood there for a moment, studying the pearls, which to my eighteen-year-old eyes looked very expensive indeed.

Then, as if on cue, the phone rang. I reached over to the desk and answered, and a man on the other end said, "I was just at your station. I had this necklace I bought for my wife and I think maybe it fell out while—"

"Hey!" I said. "I can't believe it . . . the guy's right here. He just found it in the men's room."

"That's incredible," the man on the phone said. "Look. Tell him to stay put and hang on to it. I can be there in half an hour."

"Sure thing."

"Let me give you a phone number," he said. Which he proceeded to do. "And listen . . . tell him I'm bringing $200 for his troubles. He really saved my life. Or at least my marriage!"

I put down the phone and excitedly explained to my new friend that the owner would be here in half an hour with a $200 reward.

But the guy in the station with me didn't appear too excited.

"Oh, man . . . it's not like I can wait. I gotta be in Goleta by then, and I really need to land this job." Then he looked at me and asked again, "What should we do?"

I thought about it for a moment, and he watched me think.

"I'll be here till closing," I said. "I guess I can just hold on to it till he comes."

"Would you?" He smiled brightly, then heaved a big sigh. "Man, that'd be great. So then we should split the reward."

"Really?" I said, expressing amazement, even as the wheels in my head were already churning up ways to dispose of that cash.

"Of course."

But then he bit his lip, seemingly troubled once again.

"Only problem is . . . I'm not coming back this way."

"That's okay," I said. "We can divvy it up in advance. Here . . . I can give you your half right now."

Which is what I did, actually "borrowing" $100 from the gas station's cash register, and handing it over to this guy I'd known for all of five minutes.

As I'm sure you figured out long before this point, the "pearl" necklace was paste, a cheap string of beads in an expensive-looking box, and of course the guy on the phone was in cahoots with the guy who showed up at the station.

So how could anyone be so dumb as to go along with this scam, forking over what to me was real money on the basis of such a lame story and cheesy coincidence?

Was I simply overwhelmed by greed?

Well, no doubt about it, I had dollar signs in my eyes as I looked at the jewelry and heard the magic word *reward*. But I was a reasonably smart kid, with a knack for numbers and figuring out puzzles, so if anyone should have seen through the scam . . .

It also wasn't as if I'd never been schooled in right and wrong. You think your parents were strict? Mine took me *out of* Catholic school because it wasn't *strict enough!* And although it sounds more like a punch line than the truth, before my mother was my mother, she was a nun. She had spent four years as a member of the Sisters of Loretto at the Foot of the Cross, and my upbringing, complete with Latin mass, years of breathing incense as an altar boy, and white-glove inspections of my room for dust, left no doubt that we are all born in sin and driven by base passions that have to be tightly constrained and relentlessly monitored to keep us from behaving badly. My mom's view was the classic approach to governing human nature, the top-down approach filled with "thou shalts" and "thou shalt nots" that's held sway throughout Western history. She based her child-rearing on the assumption that unselfish, moral behavior was impossible without the ever-present threat of punishment, and the more terrifying the better. So cue those images of hell from Hieronymus Bosch.

But when I think back to the incident at the ARCO station, it's not greed that I remember, or any of the other deadly sins that the philosophers and theologians (and my mother) worried so much about. I think I was motivated by a genuine desire to be of assistance. This poor guy had an important interview, and he looked flustered, down on his luck, almost desperate. With the first words out of his mouth he asked for my help, and he really looked like he needed it. But more than that, in everything he said and did, he appeared to put an amazing amount of trust in me, relying on a high school kid to get the necklace back to its rightful owner. Several times he asked me, "What should we do?" And then he left me in charge of doing it. After a show of faith like that—helping him just felt like the right thing to do.

When I went on to college, I majored in mathematical biology and economics, but questions about how we know the right thing to do stayed with me. I read a lot of moral philosophy and even theology along the way, and then after grad school, the math, the biology, the economics, and the moral concerns all came together in my early work connecting trustworthiness to prosperity.

So now let's flash-forward to November 2001.

I'm up at two in the morning lugging equipment across town and into a lab I've borrowed at UCLA by convincing a UCLA postdoc named Rob Kurzban to collaborate with me. I've commandeered a couple of graduate students to serve as Sherpas, as well as to be official passengers so I can qualify for the carpool lane on the freeway. I'm a tenured professor of economics at Claremont Graduate University, but I'm starting a very atypical research program, stretching the boundaries of my field, which means I'm now having to do science the way indie filmmakers make movies—borrowing space, begging for funding, and hauling equipment around Los Angeles in my car. We've made maybe four trips back and forth between Claremont and Westwood today, and it's at least an hour and a half each way.

I didn't know it yet, but I was about to invent a new field called neuroeconomics, and I was going to do it by running the first vampire version of something called the Trust Game.

## How the Trust Game Works

The Trust Game is a classic research tool in experimental economics, and we're going to spend quite a bit of time with it, so here's how it works. Let's say you're an undergraduate, and you need some extra cash, so you agree to take part in what's described as a study of monetary decisions. You come to a big room, like the one I'd borrowed at UCLA, along with maybe fifteen or sixteen other people you don't know, and you sit down in a small cubicle with a computer. You read the online instructions, which confirm that, just for showing up, you now have $10 on account, which is yours to keep. But soon you may receive more. That's because the computer is going to ask some other randomly chosen and anonymous player—let's call him Fred—if he would like to transfer some or all of his $10 to another anonymous player, which happens to be you.

But why would he do that? Because, according to the rules that you and Fred both just spent a few minutes reading, any amount he gives you will *triple* in value the moment it hits your account. But increasing your wealth wouldn't be entirely altruistic on Fred's part. The rules also say that if he transfers money to you, you then will be asked if you want to give some of your multiplied-by-three bonus back to him. The question is, will you? Can you be trusted to reciprocate?

The beauty of this test is that there's no social pressure to be on your best behavior because the computers mask who is doing what. Even the experimenters know the individuals only by code numbers. So Master of the Universe or Mother Teresa—the moral model you choose to follow in giving something (or nothing) back is entirely up to you. Even when you're paid at the end, no one else will know how much you made unless you tell them.

Let's say that Fred takes $2 from the initial $10 bankroll he received just for showing up, and he transfers it to you. His $2 transfer triples to $6 as soon as it hits your account, which means that you've now got $16 (10 + 6) and Fred is down to $8 (10 − 2). So you're doing pretty well. You don't know exactly who you have to thank, but you do know that you've picked up an additional $6 and that an anonymous benefactor at one of the other computer terminals in the room is responsible. You also know that your benefactor's decision was based on an expectation that you would be decent about it and share at least some of the wealth. After all, it's really no skin off your nose to flip back a couple of bucks. It seems only decent—like tipping the waitress who brings you your coffee. That's just what decent people do, right?

Let's say you decide to give $3 back to Fred. That leaves you with $13, and brings Fred up to $11—a go-ahead of $3 for you and $1 for him, which isn't much, but still better than where you both started. Then again, you're perfectly within your rights, if you so choose, to walk away with your original $10, plus the $6 bonus Fred made possible, without so much as a *Thanks, chump.*

As the amount being transferred increases, the potential payoff becomes more interesting. If Fred is really, seriously trusting (or reckless), and decides to bet the farm by giving you all of his original $10, that amount will triple into a $30 windfall, which pushes your $10 bankroll up to $40. If you're scrupulously fair-minded, you'll split the new total with your anonymous partner, and you'll both walk away with $20, or twice what you would have earned if he hadn't trusted and you hadn't lived up to that trust.

But here's the $64,000 question: If you're under no obligation to be trustworthy, and nobody knows whether you are or not, why would you ever reward trust from a stranger with a reciprocal gesture that takes real money out of your pocket? If no one's ever going to know, what's the problem with being a greedy bastard and screwing the other guy? Well, according to the economic theory that held sway over most of the twentieth century, that's exactly what you should do.

Economists had fallen in love with a concept called "rational self-interest," which assumes that each individual makes decisions on the basis of personal advantage, and also on the basis of a rational calculation as to exactly where that advantage lies.

Economic theorists had been inspired by the ideas of theoretical physics, mostly in the area of thermodynamics, with its systems of inputs and outputs moving toward equilibrium. The beauty of rational self-interest as an organizing principle was that it allowed economists to vastly simplify the math in their models. If humans always make decisions (a) rationally and (b) on the basis of self-interest, then model builders don't have to take into consideration emotions, personality quirks, or sudden flights of lunacy. Each person—or at least the theoretical person who lives inside the models—always sizes up her options and makes a logical choice based on what's best for her.

A fellow named John Nash, the subject of Ron Howard's film *A Beautiful Mind*, actually won the 1994 Nobel Prize in economics for his work refining rational self-interest into an even more elegant and hugely influential formula called the Nash Equilibrium. According to Nash's theorem, your response in the Trust Game should be simply to keep whatever comes to you, even though you know some other person increased your wealth partly in the hope that you'd reciprocate. In the same fashion, the Nash Equilibrium says that this other person should have enough sense to expect self-interested behavior from you and not trust you with a dime. After all, you've never so much as said hello. Of course, the unintended consequence of such "rational" behavior—that is, looking out for number one—is for both of you to miss the opportunity to gain by creating a larger pie, then sharing it.

For more than a century, the idea that human behavior is fundamentally both rational and self-interested was presented as gospel to millions of students, including many of those who have gone on to run our most powerful businesses and government institutions. These are the people who often set the standards for behavior on Wall Street, in government, and in the boardrooms of global corporations. Yet with all deference to John Nash and his Nobel Prize, the Trust Game shows that rational self-interest is bupkis when it comes to real people.

In the United States the stakes in the game have been as high as $1,000, and in developing countries as high as three months' average salary. With large sums or small, in dollars or dinars, participants almost always behave with more trust and trustworthiness than the established theories predict that they will. In my own experiments with the game, 90 percent of those in the A-position (the trusters, like Fred) send some money to the B-player (the recipients, like you), and about 95 percent of the B-players send some money back, based on . . . what? Gratitude? An innate sense of what's right and wrong?

Or could the behavior possibly have something to do with a reproductive hormone with curious properties involving trust and reciprocal trustworthiness?

## A Crackpot Notion?

One of my colleagues told me that this was "the stupidest idea in the world," but to me it made perfect sense. At least it made enough sense that I wanted to check it out before I dismissed it as a crackpot notion.

Our human guinea pigs—the UCLA students who'd agreed to be tested in exchange for pizza money—began to drift in and take their seats around nine thirty in the morning. At ten o'clock I got up in front of them in my spiffy new lab coat to make a few opening remarks. I thanked them for agreeing to participate, and then I reminded them—we'd explained all this in a recruiting email—that they'd already earned $10 just for showing up.

I then gave a rough overview of what we were going to do—the same story about player A and player B that I related to you a couple of pages ago—but with an added feature. Just after the decision-making, we were going to strap tourniquets around the players' arms and take their blood.

There was no visible reaction. They hardly seemed aware of me. They hardly seemed awake.

I told everyone to log into the computers in their booths using their identity-masking code, and to read the instructions. The protocol described in greater detail how their decisions could turn the $10 they'd already earned into more money, or how their decisions could cost them money.

Now I began to see some raised eyebrows and slightly more animated expressions. Everybody seemed to be waking up. It was as if they were thinking, So what is this? *Who Wants to Be a Millionaire* on a budget? Or maybe *Who Wants to Be a Millionaire* on a budget meets *General Hospital*?

I had to keep everyone occupied while we focused on each individual participant's decision and blood draw, so I asked the larger group to start filling in a personality survey.

Then I started calling out the code numbers for various players, selected in random order. "Number Six, please make your decision. And as soon as you're done, please raise your hand."

The question at this point—a question to which we thought we knew the answer—was whether or not any given A-player would choose to transfer some or all of his money to a randomly designated and anonymous B-player. Would player A trust enough to give money, counting on player B to reciprocate by giving something back?

When one of my graduate students saw a hand go up, she would immediately escort the A-player, the decider, to the smaller room off to the side that we'd set up for the blood draws. It seemed unlikely that the kind of decision put before the A-player, which was a pretty cold calculation, would affect oxytocin, but we took their blood anyway because we didn't know—no one had ever done this experiment before. What we did know was that any hormonal change in either player would be transient. Animal studies had shown that oxytocin surges in response to the right kind of stimulation, then fades after about three minutes. Which meant that the blood had to be drawn right away.

On hand to do the honors was an internist from Van Nuys named Bill Matzner. In mid-career Bill had decided to do graduate work with me, focused on health care economics. I talked him into vampire economics instead, and now he had been dragooned into being my blood tech.

As a medical doctor, Bill was invaluable to my improvised research effort—at this point, remember, I was still a whiteboard-and-computer kind of guy, not a laboratory kind of guy—donating everything from the Band-aids and cotton balls to the centrifuges, the mechanical devices that spin blood so that the serum and the plasma separate from red blood cells. But with an established practice and lots of assistants, he'd grown a bit rusty on drawing blood, so I had him practice on me. I didn't want to torture people needlessly, so we rehearsed every aspect of the protocol, endlessly, making sure that we could move fast and that nobody's time (or blood) would be wasted.

Another problem was that the centrifuges Bill had been nice enough to contribute were not the $7,000 refrigerated kind. Oxytocin not only fades fast in the body but also degrades rapidly at room temperature, so you have to grab it fast and keep it cold. Luckily, I'd been planning this new venture for a long time, and while roaming the campus at the end of the spring semester I'd stumbled upon some undergraduates packing all their stuff into their cars to go home for the summer. Without too much trouble I'd been able to talk them into donating their mini-fridges to the cause of science.

With our less-than-cutting-edge technology, we developed a protocol that involved spinning down the samples inside the cube refrigerators, transferring the separated blood products into microtubes, flash-freezing them to –100 Centigrade using dry ice, then storing everything in Bill's ultracold freezer twenty minutes from UCLA until we had a sufficient number of samples for analysis.

Once all the A-participants had made their decisions and we'd taken their blood, we allowed the computer to release the results to the B-players. A few might have been stiffed, but based on the Trust Game's history, we knew that most would have the pleasant surprise of a few extra bucks added to their bankroll.

Now it was time to see how many would be willing to split the difference and give *back* a portion of their newly acquired wealth.

"Number Nine, please make your decision. As soon as you're done, please raise your hand."

Once again, if being trusted by an A caused oxytocin in a B to spike, we had only a few minutes to capture the surge.

Participant 9 sat down and rolled up his sleeve; Bill applied the tourniquet. Then Bill jabbed the needle. Then 9 howled in pain. Bill jabbed again, and again our participant shrieked. I glanced into the main room where I could see all our test subjects turning back to look toward the sound. Apparently Bill could have used even more practice than all the sessions we'd put him through.

Another volunteer fainted, which put us on the horns of a dilemma. We didn't know how many good samples we were going to get, and with each person we had to move fast before the faint trace of oxytocin returned to baseline.

We hovered over the poor guy, Bill with his syringe, a graduate student holding our unconscious test subject as he slumped in the blood-draw chair.

"What do you want to do?" Bill asked me.

I was desperate for data. "Let's get his blood," I said. "Then we'll revive him."

But even with orange juice and cookies we still couldn't get him up and running again. I told the other participants that we'd had a glitch and that they should just surf the Web while they waited for us to resolve it. It took fifteen minutes, but we finally put our fallen comrade back on his feet.

Walking back through the room to resume the experiment, I

noticed that one of our subjects had some racy images on his computer screen—not porn, exactly, but a music site where the videos were pretty steamy. Worrying about outside-the-lab influences on him, I noted his code number when he went for his blood draw, and when I checked later sure enough, his oxytocin levels—it's a reproductive hormone, right?—were through the roof. Given the "external stimulus" he'd been receiving, we had to toss out his data.

Over the next year and a half, we repeated this vampire version of the Trust Game fourteen times. Again, this was make-it-up-as-you-go-along, gonzo science, so each experiment had to wait until I could raise a few thousand dollars in grant money, lug all our equipment in freeway traffic to UCLA, run as many sessions of the experiment as I could afford, then stash the blood products in Bill's freezer at his office in Van Nuys on the way home. Eventually we had enough samples to carry out a statistically meaningful data analysis, and then some.

This is what we found.

First, we saw the high levels of trust and trustworthiness we anticipated, the morally benign behavior that defies rational self-interest and the Nash Equilibrium. We also found significant economic rewards for virtue—which, given my work on the factors that make societies prosperous, came as no surprise. A-players who decided to trust their anonymous partners walked away with an average of $14, which was a 40 percent go-ahead over the $10 they started out with. The B-players who received money from a partner who trusted them to reciprocate left the lab with an average of $17, which was a 70 percent increase. So positive social behavior was increasing the prosperity of our little population of undergraduates, even if the benefits were not distributed with perfect equality.

But what was going on at the level of blood and brain? In this first vampire Trust Game we were truly winging it, so we had to be cautious about over-interpreting and drawing unwarranted conclusions. (Plus, I was an economist! What did I know about blood values?) That's why we kept doing the study again and again until we had a ridiculously large sample on which to base our conclusions. And what we found was a dramatic and direct correlation between a person's level of oxytocin and her willingness to respond to a sign of trust by giving back real money.

Then again, multiple factors can feed into almost any biological or behavioral response. So to pinpoint what was—and what was not—causing the virtuous behavior, we measured nine other hormones that interact with oxytocin to see if they were having any influence. These included the male hormone testosterone, as well as the female hormones estradiol and progesterone. Then we correlated all the physiological data with personality survey questions such as "Do you look through your roommate's stuff when they're gone?" and "How much do you drink?" and "How often do you go to church?"

After enough analysis to make your forehead bleed, we found no link between any of these other factors and the reciprocal generosity we were seeing. The only factor that could explain the behavior was the increase in oxytocin. But how did we know it was *trust* that was driving the oxytocin response? How could we be sure it wasn't just the receipt of money?

To check this out, we ran a control experiment in which all the circumstances were the same, except for the element of one human being's faith in another human being. Rather than have the A-player decide on his own whether or not to transfer money to B, we set up a way to make the allocation random. In keeping with my low budget, indie-filmmaker way of doing science, I drove over to Walmart and got a clear plastic container, covered the outside with duct tape, and filled it with Ping-Pong balls numbered from zero to ten. For this randomized, it's-not-about-being-trusted version of the game, I would call out an ID number, and one of our A-participants would publicly (and randomly) pull out a numbered ball. The amount on it would then be subtracted from his account, then tripled in the account of a randomly selected B-participant. The transfer of money was still taking place, but there was no human bond at the root of it.

When participants received transfers of money based on someone's decision to trust them, their oxytocin levels were 50 percent higher than the levels of those who received money based on the Walmart bucket and the random luck of the draw. Those who knew that their windfall was based on another player's faith in them also returned almost twice as much—41 percent of their new total—compared with the amount returned—25 percent—by those whose good fortune was random.

As icing on the cake, when the original transfer was based on trust, there was also a directly calibrated correspondence between the size of the transfer and the size of the recipient's response. The more money sent, the higher the oxytocin level; the higher the oxytocin level, the more money given back to player A. When the money came from a random transfer, there was no correlation at all between the level of oxytocin and how generous (or not) the B-player chose to be.

We had just discovered the first non-reproductive stimulus for oxytocin release in humans. Which made me very happy for a variety of reasons, some of which involved my frustrations with the profession in which I'd been working.

## The Forgotten Bond

In its "physics envy," mainstream economics had embraced mathematics to the neglect of any real interest in human nature. This, despite the fact that economics actually came into being as an offshoot of moral philosophy. And the central question of moral philosophy—whether human beings are fundamentally good or evil—has to be the longest-running debate since debates began.

Not too long after Moses picked up the Ten Commandments on Mount Sinai, the Psalms described humanity as being "a little lower than the angels." Arguing for the other side, the Roman playwright Plautus declared that "man is wolf to man." Philosophers, preachers, and politicians have been going at it ever since, offering theories to pin down our moral core that range from the medieval idea of original sin, to the seventeenth-century idea that our natural state is "the war of all against all," to the Romantic idea that we are born a blank slate upon which all manner of goodness might be written if only we have the right environment in early childhood.

And this is not merely some academic dispute. This is a debate with consequences because each contending theory competes for influence in our laws, our cultural norms, and our social policies.

Two hundred and fifty years ago, an obscure professor at the obscure University of Glasgow published a book called *The Theory of Moral Sentiments,* arguing that benign and generous behavior arises from our feelings of attachment to others. He said that seeing others in distress creates a bond that he called "mutual sympathy."

In hindsight, this seems almost self-evident. We know that seeing others in distress can have such an immediate force that it makes soldiers throw themselves onto a grenade to shield their buddies from the blast. Sometimes it compels ordinary people to jump down onto the subway tracks to save a complete stranger from being crushed by an oncoming train.

Yet *The Theory of Moral Sentiments* created such a stir that students from all over Europe suddenly flocked to Glasgow to study with its author. Overnight the obscure professor became one of the intellectual rock stars of the eighteenth century, even though, with bulging eyes and neurotic twitches, he hardly fit the part. He lived with his mother, and he was so absentminded that he often got lost in the woods, talking to himself, dressed only in his underwear. Still, the concept of mutual sympathy was such a bolt from the blue, and his book such a hit, that he was able to travel grandly and lecture and hobnob with the likes of Voltaire and Benjamin Franklin for the rest of his days.

So what was all the fuss about? Well, for centuries, most moral thinking was like my mother's, bound up in original sin and the fall of Adam. But here was a theory to explain moral behavior that was not all about reining in our "natural" depravity. This theory did not assume, like the seventeenth-century philosopher Thomas Hobbes, that our natural state was "the war of all against all"; nor did it rely on a higher authority, or on a mystical sixth sense, or on rational calculation and restraint to help us overcome our dog-eat-dog proclivities. Instead, *The Theory of Moral Sentiments* suggested that conscience and good behavior are inherent parts of our psychological makeup, and that they are elicited quite naturally from our social relationships. Discerning right from wrong is, in other words, an innate human ability, and a bottom-up response from deep within.

Most secular philosophers had maintained something very much akin to the church's dismal view of our natural inclinations as well as a similarly top-down approach to getting us to shape up. The only difference was that instead of the God of Wrath threatening us into submission, the top-down force that philosophers saw struggling to impose control was human reason. Plato described the mind as a charioteer trying to rein in the body's wild, animal impulses, which he characterized as spirited horses. A couple of thousand years later, Pure Reason picked up an even more zealous advocate in the person of German philosopher Immanuel Kant.

In Kant's view, the only thing that makes us human and free is to act in accordance with the rules we give ourselves, devised through reason. The most fundamental of these rules, what he called the Categorical Imperative, says that to arrive at the good, you must always act as you would if your action were to become a universal law. But where the purity of Kant's Pure Reason may have jumped the rails was in saying that for any action to be truly moral, it must be done *entirely* for the sake of the moral law. If we act morally because it *feels good* to be virtuous—that doesn't count. And no exceptions, regardless of outcome. If lying violates the universal law, then you absolutely must never lie, even if a psycho killer is after your friend and telling the truth about his whereabouts will lead to his death.

If this line of pure reasoning seems a bit cold and impractical, that's only one of the many problems with top-down approaches in general. The ones that, like my mother's, rely on religious teachings bump into the obvious fact that there are somewhere around four thousand different religions in the world, each adding its own special rules to the basic guidelines for pro-social behavior. Throughout history, nothing has led to more bloodshed and ruthless brutality than conflicts among these differing approaches to God. Which is precisely why the secular philosophers tried to rise above all that discord and find universal answers through reason. But in that effort, the philosophers carried over the same contempt for our biology that often characterizes religion. The effort to leave "mere flesh" behind depends on the notion that the mind—and the will, and the soul, and the indomitable human spirit—somehow stands apart from the body. Which is a view that modern science has proved—sorry, Mr. Kant—to be just plain wrong.

We are biological creatures, so everything we are emerges from a biological process. Biology, through natural selection, rewards and encourages behaviors that are adaptive, meaning that they contribute to health and survival in a way that produces the greatest number of descendants going forward. Oddly enough, by following that survival-of-the-fittest directive, nature arrives at many of the same moral conclusions offered by religion, namely, that it is often best to behave in a way that is cooperative and, for want of a better word, moral. Nature simply gets to the same place by following a different, and perhaps more universal, path.

The notion of mutual sympathy was much more human-centric than anything that had come before, just the kind of moral philosophy that the budding Romantic Movement, poised to give the world the Noble Savage and *The Rights of Man*, could get behind in a big way. If much of human history appeared driven by the ruthlessness that obsessed thinkers like Hobbes, perhaps it was because of specific influences on the system. Alter the nature and extent of those influences, and you might alter the moral response.

The eighteenth century was still a very long time before science could contribute much to a discussion of behavior, so our nerdy professor from Glasgow was understandably a little vague about how this system of mutual sympathy operated. Still, we see something very much like it—we call it empathy—driving moral conduct in thousands of little kindnesses every day. Every day, all over the world, it compels billions of people to share what they have with others whom they care about.

And yet, after the initial surge of enthusiasm, mutual sympathy lost the battle of the big ideas in moral philosophy. In part, it was overshadowed by Kant's ideas about Pure Reason, offered at about the same time. But there was another intellectual hammer coming on the scene with even greater impact.

Romanticism may have captured the arts and, to some extent, the politics of the late eighteenth century, but in the workaday world, the real spirit of the age was a new idea called Capitalism. Enterprise was on the rise, and tradition was in decline. Men of wealth and power were forming trading companies and building factories, meanwhile dispensing with medieval ideas like the fair price and noblesse oblige. Once their big machines were ready to roll, they closed the open grazing land so that tenant farmers would have no choice but to go to work in the mills.

The man that Capitalism turned to for hardheaded, unsentimental moral guidance to this new age of enterprise was Adam Smith, author of *The Wealth of Nations*. The irony is that Adam Smith is also the same head-in-the-clouds professor whose first book had put human feeling at the center of moral discourse. It was, in fact, the leisure he earned by way of *Moral Sentiments'* success that enabled him to write *The Wealth of Nations*, which had an impact that, by comparison, makes *Moral Sentiments* look like a dud.

Many factors account for its electric effect, but one sentence, quoted time and time again over the past two centuries, conveys the basic wallop:

> It is not from the benevolence of the butcher, the brewer, or the baker that we expect our dinner, but from their regard to their own interest.

At a time when the West was moving beyond ideas of sin and the limits those ideas imposed, here was a real game changer. In the medieval world, pursuing personal gain fell under the rubric of pride or envy or greed. But now, according to the already rock-star-famous Adam Smith, personal gain could be filed under a new linguistic category called one's "interest," and it was not a vice at all but a virtue! Getting ahead was no longer seen as a result of unruly passions. Now in the Age of Reason, getting ahead was simply the *reasonable* thing to do. And best of all, the rational and reasonable pursuit of personal gain made the wheels go around that put more food on the table for everyone.

Ever since those fateful words of Smith's first appeared, what's been lost in the general enthusiasm for the self-interested butcher and baker is how that line of text fits within the context of Smith's larger intellectual enterprise, which had far more to do with the virtue of individual initiative than any endorsement of self-serving behavior. All the same, Smith was embraced and venerated as the founder of a new science called economics. At the same time, his status as a moral thinker fell into decline.

For economists, Smith's line about "their own interest" represented not only a shift in values, but the possibility of a new and comprehensive way of explaining behavior. And that's how Economic Man (also known as *Homo economicus*) was born, the highly rational, self-serving human who lives in economics textbooks and in economic models, and who—at least for theoretical purposes—is driven by anything but mutual sympathy.

As an economist who went on to study moral behavior, I've always had a soft spot for Smith, the misunderstood moralist who went on to found economics. Like him, I've always preferred to study actual *Homo sapiens* rather than theoretical *Homo economicus*. I was always drawn to the real-world underpinnings of economic issues—things like rates of childbirth, generational demographics, and the amount of resources parents invest in each child. Don't all parents love their children? Then why don't all parents express that love by trying to give their kids the best possible preparation for life? Usually it's because they don't have the time and resources, and that's usually because they have more kids than they can handle. It turns out that fertility and parental investment—biological issues—profoundly affect economic outcomes.

It was this kind of work on fertility and demographics that prompted me to investigate other interpersonal factors affecting prosperity, the most compelling of which was trust. I spent more than a year developing my model demonstrating that the level of trust in a society is the single most powerful determinant of whether that society prospers or remains mired in poverty. Being able to enforce contracts, being able to rely on others to deliver what they promise and not cheat or steal, is a more powerful factor in a country's economic development than education, access to resources—anything.

## Enter Oxytocin

In 2000 I attended a conference on economics and law held by the Gruter Institute for Law and Behavioral Research. This was in the summer off-season up at a ski resort in the Sierra Nevadas, and on the long shuttle ride from the Reno airport I found myself seated next to the only other passenger not decked out for a mountain biking trip. We got to talking—this other passenger was indeed heading to the same conference—and that is how I met anthropologist Helen Fisher, author of such books as *The Anatomy of Love* and *Why We Love*. We started comparing notes on our research, and I mentioned my studies of parental investment, and after a while she asked me, "Have you ever thought about studying oxytocin as a factor in all this?"

Oxytocin? I'd never heard of it. But when she described it as a bonding chemical, I took the bait.

Later, back in my hotel room, I went on PubMed and soon learned that oxytocin is a small molecule, or "peptide," that serves as both a neurotransmitter, sending signals within the brain, and a hormone, carrying messages in the bloodstream. In 1906, when Sir Henry Dale first identified it in the pituitary gland, he gave it a name by combining the Greek words for "quick" and "childbirth." Obstetricians and gynecologists came to know it well because it controlled the onset of labor and the flow of milk for breast-feeding. But beyond the realm of reproduction, researchers in medicine apparently never gave it a thought.

I was intrigued, though, especially when I found a large body of oxytocin research done by the kind of biologists who studied small, furry animals. Injected directly into the brain of some species (not allowed with humans, by the way), oxytocin worked like a mythical love potion, creating an instant and powerful monogamous attachment. In the highly social world of moles, voles, and prairie dogs, it was shown to regulate all forms of attachment, including bonding to a mate, tolerance of neighbors in the cage or colony, even tolerance of one's own offspring. By inhibiting oxytocin, researchers had induced mothers to shun their offspring; when other scientists induced the release of oxytocin, it caused mothers to nurture offspring not their own, just as nursing dogs occasionally adopt orphaned kittens.

And it was the off-and-on quality of the hormone that intrigued me even more. In nature, oxytocin surges when signals from the environment indicate that it's safe to relax and nuzzle. When those signals wear off, or when they're countermanded by some other signal—such as danger—it's time to get back in the game of gnashing teeth and competing over resources.

Reading about all this research in the biology journals, I couldn't help thinking that the oxytocin signal—a calm but transient feeling, heavily dependent on an assessment of safety in the moment—sounded a lot like trust. And that's when the really interesting possibilities began to tumble out. Bonding . . . trust . . . parental investment . . . These seemed like entirely different concepts, until you thought about the underlying mechanism.

What if bonding in voles and trust in humans were actually based on the same chemistry? What if oxytocin was, in fact, the chemical signature for that elusive bonding force Smith had called mutual sympathy? Then, thinking back to my research on the prosperity-enhancing power of trust, I had to laugh. What if this "Moral Molecule"—if that's what oxytocin was—is also an essential element in what Smith called the wealth of nations?

This was a eureka moment for me, where the possibility of so many ideas coming together made me a little giddy. If I could demonstrate a direct link in humans between oxytocin and concern for others, then it would mean that this notion of mutual sympathy was not just an abstraction or a pre-scientific metaphor like "the four humors." I could well imagine that, with a few million years of evolutionary refinement added on, the same basic system that allowed primitive creatures to let down their guard and mingle, then resume wariness when it was time, could help modern humans walk the line between competition and cooperation, benevolence and hostility, maybe even what we call good and evil. And given that trust was the number-one factor in helping societies move toward greater prosperity . . .

Well, it was one hell of a theory, but a theory doesn't get you much unless you can prove it. So that's when I started retooling to add blood and brain work to my portfolio of research techniques. Spending time in my father's engineering lab as a kid, I'd learned the value of tinkering and exploring outside the usual boundaries. So I went back for training in neuroscience at Massachusetts General Hospital. I started hanging out in the neurology department at the nearby medical school, attending lectures and grand rounds. I was already a full professor—but in economics, not neuroscience. So this new interest of mine meant starting over.

This was about the time that I mentioned my new research program to a friend who was an ob-gyn. "That's the stupidest idea in the world," he told me. "It's a *female* hormone."

"So what? And by the way . . . men make it, too."

"But it's trivial. Parturition. Lactation. That's all it does."

I simply had to trust my instincts. If I was wrong, I knew that at the very least I was *testably* wrong, which meant we'd get an answer, yes or no.

Eventually I wound up with a huge freezer filled with blood in my academic office in the economics department. This prompted a dean of mine to refer to what I was doing as "vampire economics," but I didn't mind taking some razzing. I was determined to find out if this idea of mutual sympathy had any real substance, and the only way to do that was to get down under the skin. Which is what we did, beginning with those Trust Games at UCLA.

What had once seemed like such a stupid idea—having benign, pro-social behavior being triggered by a reproductive hormone responding to trust—now seemed too good to be true, almost like some scientific version of a parable you'd learn in Sunday school. Then again, if oxytocin allowed voles to get along better with one another in their colonies, why not humans? For social species, if moral behavior is more adaptive than ruthless behavior, then it only made sense that there would be a biological basis for it. And where would it be more likely to originate than in reproduction, where all bonds and attachments begin?

I thought about this biological imperative as I watched our volunteers leave the room that first morning at UCLA. They had to stop by a cashier to pick up the money they had earned, and—this being a population of young, single undergraduates—there was a fair amount of heat being generated as the kids checked out one another.

I listened in on their conversations, which included a lot of "Which one were you? What'd you do? How'd you make out?"

Not surprisingly, I didn't hear a single guy say, *I was a total bastard. I took everything I could get and gave absolutely nothing back.*

Nor did I hear a single girl say, *Yeah. I tend to be cold and withholding, and I don't really trust anybody. So I just kept my ten bucks. Screw 'em.*

Based on the self-reports I heard, you'd think everyone in the room was applying to Teach For America, helping out in soup kitchens, and reading to the blind. Every student I heard claimed to have been a shining example of moral probity, either magnanimously trusting or generously trustworthy.

Which prompted two additional observations.

The first is that pro-social behavior is a sexual come-on. In fact, gift-giving—the display of generosity—is rule number one for courtship in all human societies, and in many animal ones. Who wants a mate who's going to be selfish and self-interested?

The second observation is that people will lie like crazy to impress a potential mate. Then again, human beings are extraordinarily good at identifying liars. To make a claim of being trustworthy credible over time—as opposed to, say, a con man's brief encounter—one really has to be trustworthy.

So it makes sense that nature would go with that old Russian adage: "Trust, but verify." Oxytocin is the touch-and-go molecule that makes it possible to walk such a fine line: Trust and bond with someone when the right stimuli are present, but be prepared to return to wariness once the stimulus fades. How oxytocin came to be that carefully modulated governor of trusting behavior, and how trust led the way to more complex social behaviors such as empathy, is a much richer story—one that takes us back in time, and into the deep blue sea.

# CHAPTER 2

# Lobsters in Love

*The Evolution of Trust*

Consider, if you will, the lowly lobster. The scary-looking crustacean *Homarus americanus* has never been thought of as particularly moral, or particularly romantic (unless, of course, the context is drawn butter and a good white wine). Heavily armored and heavily clawed, these beasts are highly aggressive, highly territorial, and, at least in captivity, known to dine on one another.

But when the mood is right and the lights get low, lobsters actually can be kind of sweet, with a courtship ritual that's like a soft-focus scene from an old French film. It all begins when the female sprays a seductive perfume into the male's grotto, then scuttles inside to slip out of her shell. Just as in so many movie scripts, finding the right mate requires letting go of that tough, protective exoskeleton. But for a lobster, leaving the shell behind means being utterly vulnerable until a new one grows back. Which suggests a huge leap of faith. The female must trust her life entirely to the male she's chosen, a creature she would ordinarily treat as a competitor, if not an outright threat. The chemical signal that allows her to suspend her wariness just long enough for the tryst, and for the growth of a new shell, is an ancient precursor of oxytocin. A related chemical that will show up later in our morality tale prompts the male to watch over her, protect her, and treat her gently.

Can we call what we see in lobster courtship "trust," and moral behavior in response to trust? That would be getting ahead of ourselves by about a hundred million years. What we can say, though, is that the most basic, physiological mechanism for all our moral impulses dates back to a time long before animals ever ventured onto dry land. And it all began with sex.

The fact that the precursors of trust and of reciprocity are so primal, that the ancestral DNA of our moral behavior is embedded in cells throughout the body, and that it is all rooted in reproduction, suggests pretty clearly that what we now call morality is not some civilizing afterthought, or a frill that runs counter to nature, but, in fact, something deeply connected with basic survival.

If the biology of reproduction seems like a lowly and unlikely starting point for the kinds of lofty issues that would later concern prophets and priests and philosophers, ask yourself which would bother you more: that your spouse fudged a little bit on last week's expense account, or that he or she had a little extramarital sex during an unexpected stopover in Dallas? The desire to steer sexual energy toward the most socially constructive outcome is very near the core of every moral system in every culture on the planet.

This chapter offers a sort of Nature Channel overview of how this system of moral guidance, grounded in the chemistry of reproduction, originated and evolved. First it established bonds between mates, then between that mating pair and their offspring, and then between the members of that nuclear family and their immediate kin and companions. How we came to know what we know about how this all works is a remarkable detective story.

But the answer to this whodunit gets us into the most important question yet to be resolved by us as a species. Under the influence of oxytocin, it's not too hard for us to behave with generosity, care, and concern toward those with whom we share a deeply personal bond. The bigger challenge is this: How do we extend that kind of virtuous behavior to those with whom we have almost nothing in common, and with whom we will never have a face-to-face encounter?

To begin to answer that question, we need to get a running start with an evolutionary story that goes back about seven hundred million years. The first characters we meet in this tale are sea creatures so primitive that their nervous systems operate more like computer code than what we'd think of as a brain. For computers, the choice is always binary, meaning that there are only two options. For these ancient beasts, the binary choice was not between a one and a zero but between yes and no, stop or go, approach or withdraw. A hunger impulse would prompt an advance. A harsh or painful stimulus would prompt a withdrawal. A threat would stimulate stress hormones that would prompt either a withdrawal or a show of hostility—the famous fight-or-flight response. Mental anguish over moral ambiguity—Should I cheat on my husband who's been cheating on me? Steal from the tax-evading company that employs me? Kill one person to save five others? Appease Hitler to avoid a war?—was simply not part of the package.

The clusters of chemicals that orchestrated these stop/go responses operated like rocket thrusters guiding a spaceship. Working in opposing teams, they turned on and off, pushing our most ancient ancestors in opposite directions at different times. If the animal was programmed correctly, it would zig when it should zig and zag when it should zag, and it might just stay alive long enough to reproduce. Which is how nature always measures success—in terms of surviving long enough to pass your genes down into the next generation, and then the next.

In the second act of this story, however, the need arose for behavior that was more accommodating and subtle than the all-or-nothing of approach/withdraw. More complex animals such as the lobster actually have to get together to mate, but the anxiety-driven fight-or-flight response that's always made animals wary of one another was too valuable to set aside. After all, wariness also helps animals survive. What was called for, then, was a *temporary* suspension, triggered by the right circumstances, a kind of truce that would last just long enough for courtship and mating, then fade when the tryst was over.

Whenever innovation occurs in nature it's stumbled upon by accident, and in the tiniest of increments over ridiculously long periods of time. In living systems the primary mechanism of change relies on the genetic mistakes known as mutations. When one of these accidental innovations works better than what existed before, the new gene persists and spreads. It is "naturally selected" by its own success, meaning that it keeps a particular kind of creature alive longer, in greater numbers, and producing viable offspring in greater numbers. But because the whole process is trial and error, new systems get built on top of old ones, with the new instructions written over the old in upgrades rather than replacements.

Back in the ancient seas we're talking about, at a time when all animal life was marine life, the primary stress hormone was a fight-or-flight chemical called vasotocin, which happens to be made up of nine amino acids. One fine day, quite by accident, some

31

long-forgotten fish came into the world with two of those nine acids swapped out. This new protein with the two-acid difference—we now call it isotocin—produced an effect that was the exact opposite of stress, or fight-or-flight. Instead this new molecule temporarily reduced anxiety, which allowed an animal to relax, which reduced the fear of an encounter, which facilitated sex, which proved to be a good thing. Which is why this mutant protein hung around and proliferated, becoming, over time, a standard feature in the body chemistry of fish. With isotocin, the old repertoire of approach-or-withdraw expanded to include mutual approach. Isotocin also added one more vitally important option to fight-or-flight, namely, "frolic."

Through millions of additional mutations over hundreds of millions of years, isotocin and vasotocin continued to evolve as nature made its haphazard way toward higher forms of life and, eventually, to you and me. Over time, a variant of isotocin morphed into oxytocin. Vasotocin became arginine vasopressin, and today, working together, they still serve as two of the primary "rocket thrusters" that guide our own reproductive—and moral—behavior.

We still have a long way to go before we get to the scene with the guys in white robes discussing virtue in the brilliant sunlight of Athens, but even so the moral dimension of the oxytocin family of molecules began to emerge well back in time. The trigger point was when the reproductive plot thickened with the addition of another element that has driven males and females crazy ever since: choice.

Male guppies prefer big females because big females produce more offspring, a bias that's self-perpetuating simply because more male guppies are born with the genetic preference for big females inherited from their dads. Females from most species prefer males that are strong and fierce and dominant. This, too, is self-perpetuating, because the macho genes from those macho males allow the mother's offspring to have more reproductive success themselves.

When this element of choice, called sexual selection, was added to basic competition for survival called natural selection, the pace of evolution zoomed into fast-forward. Females do most of the choosing in this ramped-up game, and as every guy knows all too well, female choice puts males to work trying to prove their worthiness. Behaviors that win lady fair can be simple demonstrations of fitness—witness head-butting elk, or frogs competing to see who can croak the loudest and longest. But in species where males cooperate in raising the young, the competition began to incorporate more practical considerations. In these species, the male has to demonstrate not only proof of resources for protecting and provisioning but also proof of commitment. Which is where the question of virtue first entered the picture.

Today males of the species *Homo sapiens* are expected to give their prospective mate a shiny hunk of rock that isn't good for much, except that it represents serious intent, evidenced by the ability (and the willingness) to blow three months' pay on a gesture. You can find the same pattern among birds, or in the African beetles that feed on the droppings of elephants. Here the male shows that he is reliable and committed not by heading to the jeweler, but by rolling up a "nuptial ball" of elephant dung, which he presents to his beloved, and into which she then lays her eggs.

In human courtship, generosity has been a come-on since the days when it meant sharing the grubs or the mastodon meat. Today women still like it when a guy is good to her grandmother and is willing to play video games with her kid brother because it suggests that he'll be a kind and generous husband.

But generosity also plays into an even deeper evolutionary

tradition, which is females being turned on by male traits that are actually handicaps, which are actually, in a backhanded way, displays of fitness. The extravagant plumage of certain birds says, *If I can survive in the wild with these stupid clumps of bright red feathers sticking out of my head and tail, I must be one hell of a guy.*

Generosity can be looked at as a handicap because it limits how ruthless a person will be in getting what he wants. In courtship, however, it says, *I don't have to be a greedy bastard, hogging everything for myself. I'm so fit and capable, I can afford to give stuff away! Obviously, I'll be good to you!*

Of course there's a huge gap between spending a few hours rolling up a nice ball of elephant dung and being counted on to spend the next forty years working down at the toothpaste factory, coaching Little League, taking out the trash, and being good to the in-laws. The great leap forward in this progression occurred when nature came up with a new way of life, and a new life category—the category of furry creatures known as social mammals.

## Sex and the Single Vole

Before we began our Trust Game studies with humans, the most sophisticated research on oxytocin and what might be loosely termed moral behavior was done on voles—chubby rodents with small eyes and big ears that make them look like characters in a Disney movie. In the 1980s a young scientist named Sue Carter wanted to explore how brain chemistry differed in two flavors of social creatures that were closely related but conducted themselves very differently. When she mentioned what she was looking for to Lowell Getz, a field biologist in her department at University of Illinois at Urbana-Champaign, he waved out the window and said, essentially, have at it. The prairies just beyond the campus were overrun with just the critters she needed.

The prairie vole and its cousin the meadow vole are a study in contrasts, even though they live in similar underground warrens, eat similar foods—mostly grasses—face similar predators, and share a common genetic ancestor. Actually, the females of both species conduct themselves pretty much the same. It's among the males that the behavioral differences are night and day.

Male prairie voles (*Microtus ochrogaster*) are stand-up guys. They live peacefully in social groups, remain with their mates for life, and put in considerable time caring for their young.

Male meadow voles (*M. pennsylvanicus*), on the other hand, are loners and players. They don't get along well with their neighbors, they seduce any female they can, and they move on to their next partner as quickly as possible with no regard for the patter of little feet.

A psychiatrist named Cort Pedersen had already shown that it was the release of oxytocin that accounted for maternal behavior in lab animals. Virgin females in the species of white rat that scientists typically use in their studies will attack or ignore pups they come across because there's no such thing as a maternal instinct without the right body chemistry. But prime these wary virgins with estrogen, then give them injections of oxytocin, and they will adopt any offspring that happen to be nearby—maternal instinct on the rampage. These white rat mama grizzlies attempt to suckle the babies, lick and groom them, even defend them against their actual mothers.

When these same animals become mothers and pick up the right chemicals in the natural course of things, they spend endless hours nurturing. Under the spell of oxytocin, they feel less pain and are less subject to distraction, carrying on their maternal duties even when researchers try to drive them crazy with noise and lights. But give such a mother lab rat a drug that blocks the action of oxytocin and she'll neglect her pups so completely, they'll die. Sad to say, this is the same effect we see in human mothers on crack, or women so abused that their stress hormones block the oxytocin effect.

Sue Carter's work was especially provocative because it took the link between oxytocin and reproductive behavior and expanded it to include social behavior in general. She showed that it's the number of oxytocin receptors lining the "reward" areas of the brain that accounts for how the gregarious and monogamous prairie vole conducts his entire life, and how his anti-social and unreliable cousin the meadow vole conducts his.

Reward areas activate when an individual encounters something pleasurable, whether it's food, sex, cocaine, or (in the case of humans) hearing someone on talk radio spout ideas they happen to agree with. For male prairie voles, living with a familiar female triggers the release of oxytocin, which registers in these areas, which then reinforce the behavior by releasing other feel-good chemicals, which induces these guys to settle down. This same cascade of feel-good chemicals occurs when male prairie voles encounter their offspring, which means that the brain rewards and reinforces the complete domesticity-and-fatherhood package.

Just like a good human dad piled up on the couch with the wife and kids, the male prairie vole simply feels good spending time with his mate and his young. And like the guy who gets along with everybody down at the hardware store and the VFW hall, this same male prairie vole gets the warm glow of companionship, rather than a feeling of threat, whenever he steps out into the burrow at large. Which gives the community of prairie voles a pretty mellow vibe and, in human terms, an abundance of civic virtue.

Meadow voles, on the other hand, lack the oxytocin receptors necessary to pick up the pleasure signals triggered by any of these social stimuli. This makes them like the young stud driving a pimped-out Trans Am, with a long string of ex-girlfriends but no real friends; or the neighborhood crank who lives alone and threatens to shoot anyone who steps on his lawn.

But oxytocin also activates brain systems that damp down pain and fear. Mammalian sex can be fairly rough-and-tumble, so oxytocin's ability to reduce a creature's sensitivity to pain— especially the female's sensitivity—contributes to *la dolce vita*. That's why the genes for oxytocin came out winners in the natural selection sweepstakes, the big contest that measures how many offspring you produce that live long enough to go on reproducing.

Sex requires risk-taking as well, and oxytocin provides further advantages by reducing sensitivity to psychological stresses, including fear of the unknown. Animals infused with oxytocin in the lab are significantly less cautious and more curious than similar animals without the infusion. Take a bunch of male prairie voles, or females of any mammal species, goose them with oxytocin, and they're ready to work the room, exploring and mingling. Aggression declines while closeness and cooperative behaviors such as food-sharing increase.

In socially monogamous species, oxytocin released during sex creates a lifelong bond. On the male side of the equation, though, monogamy involves more than just wanting to be around your one and only. Maintaining the pair bond means being ready to fend off other males who find your mate irresistible. So when monogamous male mammals have sex, the brain releases not only the "cuddle" molecule oxytocin but also its cousin vasopressin, the chemical that descended from the stress hormone in fish. The guard-and-defend part of fight-or-flight became a part of "frolic," with the protective behavior focused not just on mates but on offspring as well.

So whether we're talking about bumping and grinding in the night, or bumping into one another in the burrow, or bumping Junior up and down on Daddy's knee, it's oxytocin and its chemical partners making everyone more pro-social.

In humans the brain regions associated with emotions and social behaviors—namely, the amygdala, hypothalamus, subgenual cortex, and olfactory bulb—are densely lined with oxytocin receptors. But the oxytocin effect registers throughout the body, especially when the hormone binds to receptors in the heart and vagus nerve, which innervates the heart and gut, reducing anxiety and blood pressure, and giving cheeks the warm glow we associate with sex.

But there's an even larger cascade of chemicals taking place. When a positive social stimulus prompts the release of oxytocin, the Moral Molecule in turn triggers the release of two other feel-good neurotransmitters: dopamine and serotonin. Serotonin reduces anxiety and has a positive effect on mood. Dopamine is associated with goal-directed behaviors, drive, and reinforcement learning. It motivates creatures to seek things that are rewarding, and it makes it feel good to keep doing those things.

This cascade of reinforcement gets more involved, and more interesting, but here let's stick with the basic biology as well as the story of how our understanding of that biology reached the next level.

## Manipulating Morality

Natural history is like the Nature Channel—something you watch. You observe closely and take good notes and over long periods of time you begin to see how things work. But after a while, you still need to head to the lab for what are called controlled experiments, where you can isolate and manipulate various parts of the puzzle in order to test your assumptions and really nail down proof.

In 2000 an ingenious and talented neuroscientist at Emory University in Atlanta named Larry Young did a series of manipulations that showed the workings of oxytocin with amazing precision. He created an oxytocin "knockout mouse," which wasn't a tiny guy with whiskers and boxing gloves, but a lab animal in which the gene for the Moral Molecule had been knocked out. With oxytocin taken out of its genetic program, the knockout mice developed social amnesia. The loss of only this one gene and the one hormone it produced wiped out the ability to recognize other mice who'd been longtime pals. Mice that once tolerated familiar cage mates and even congregated together become cranks and loners.

Removing the behavior by removing only one gene was a pretty compelling demonstration of cause and effect. But to really nail it down, Young and his colleagues went one step further and reversed the change. They squirted oxytocin into the brains of their mutant, knockout mice, replacing the single ingredient they'd previously removed. Just like that, social amnesia disappeared. The mice recognized their old compadres and began to mingle once again.

Young next took virgin male meadow voles—the ones that are loners and players to start with—and using a harmless virus as a

vector into their brains, inserted the gene that codes for the vasopressin receptor. He then lined up a couple of gorgeous female
voles and paraded them past the boys who'd been modified. Not
surprisingly, these roguish meadow voles were hunks of burning
love. But after sex, these animals that should have been caddish
playboys now wanted to cuddle and nest and mate with the same
female again and again. Even in the presence of other perfectly
attractive females, these modified males ignored the siren song
and remained true blue. The introduction of vasopressin receptors
alone changed these rolling stones into househusbands.

Some women I know get excited when I mention genetic
manipulations that can transform louts into monogamous mates.
But every species is different in terms of its oxytocin and vasopressin receptors. (And messing with a guy's genes would never fly
with the ethics review panel at my university.)

Even so, I wanted to try something similar in my human test
subjects. I wanted to see if we could go beyond the association
between oxytocin and generosity we'd already demonstrated in
the Trust Game and use manipulations to remove any doubt about
what was actually causing what. All I needed was a way of putting
oxytocin directly into the human brain. I could then have test subjects play the Trust Game, and if all went well I would be able to
flip on the oxytocin like a light switch. If I also made sure that all
the other variables that might affect behavior remained the same,
this would be an absolutely clear comparison between "your brain
on oxytocin" and your brain without the pro-social surge.

Years ago, doctors created inhalers that could administer a pre-
measured dose of synthetic oxytocin to new mothers who needed
a little help with breast-feeding. But this kind of infusion isn't like
a spritz of Sudafed to clear the sinuses. It takes four puffs in each
nostril to upload a teaspoon of active ingredient into the system,

and all that snorting and dripping is not a pleasant experience. When I set out to adapt this technique for our study, though, the much bigger problem I ran into was red tape.

In the 1980s the Food and Drug Administration had approved an oxytocin inhaler for use by mothers in the United States, but it flopped in the market, so this FDA-approved version was no longer manufactured. This looked like a pretty significant setback for my scheme, but then a Swiss psychologist named Markus Heinrichs sent me a copy of his dissertation. He'd infused oxytocin into human subjects to look at its effect on stress. So evidently, oxytocin inhalers were still available in Europe.

I called Markus and he directed me to a pharmacy in Switzerland where, with the appropriate prescription, I could order an oxytocin inhaler packed in ice and shipped to me in California. Suddenly I was back in business, but then I started filling out the paperwork. For the FDA to approve these imported inhalers, trade name Syntocinon, for use in the United States, I had to show that the formulation was identical to the oxytocin formulation the FDA had originally approved.

So now I had to track down Syntocinon's European manufacturers. This particular product had bounced around among companies several times, but with some crack investigative work I discovered that the Syntocinon inhaler was being produced by Novartis. It took me another several months, but eventually I found the person at the company who could tell me exactly what, other than oxytocin, was in the liquid they were sending up women's noses. Unfortunately, while the active ingredient, oxytocin, remained the same, there were differences in the buffers and fragrances.

This seemed pretty trivial, but the fine print is what the FDA is there to worry about. It makes the rules, and researchers have to follow those rules explicitly when working with human subjects. So I compiled all the information and sent it to the FDA. And then I sent it again. And again.

For nearly two years I kept sending my request. Eventually I reached a federal employee who could walk me through a close comparison of all the ingredients in Syntocinon and in the U.S. inhaler that had been previously approved.

But I didn't want the research to be at a complete standstill while all this was going one. According to ethics guidelines, there was still one person I could experiment on—myself.

Oxytocin quickly breaks down in the stomach, with no biological effect, so one question I needed to figure out was the rate at which I could pack the stuff into the sinus cavities so it would be absorbed into the brain rather than flow down the throat. I also wanted to see if the infusion would irritate the nasal passages, or eyes, or anything else. So I tried everything short of a drill to get oxytocin into my brain: inhalers of various types, eyedroppers for squeezing it into my nose, even snorting it from a spoon.

Self-experimentation is a tricky business, so each time I tried a new method, I would sit in my wife's office at the medical center. She's a neurologist, and she'd observe me as she went about her business. One of the most common side effects of oxytocin in women is for the breasts to leak, but I wasn't too concerned about this. As a reproductive hormone, however, oxytocin has receptors in *all* the associated body parts, including the heart. Worst case— my heart would slow so much that I would pass out, which is why it was good to be just down the hall from the emergency room.

Sure enough, I'd take in as much of the hormone as I could stand, then I'd sit in my wife's office quietly working on email, waiting to see what would happen. Every half hour or so she'd come over, rub my shoulder to see if I was conscious, and say,

"How you doin', hon?" And every time, a certain part of my anatomy would stand and salute. "Note to file: Warn male participants in future infusion studies of non-harmful but potentially embarrassing side effects."

After a two-year wait, my contact at the FDA informed me that the American and European versions of the oxytocin formula were not a match.

"So where do I go from here?" I asked him.

"Clinical trials," he said. "Stage One through Stage Four. There's no other way to prove it's safe."

That was absurd, and now I was getting desperate. Clinical trials are what pharmaceutical companies do when they're going to launch some new drug that will bring in billions in the marketplace. They take years and are hugely expensive.

Luckily, as it turned out, a feature story was just about to appear in the *New Scientist* (written by Linda Geddes of Vampire Wedding fame) describing my initial Trust Game experiments with oxytocin. The editor asked me if he could send proofs of the article to an experimental economist in Switzerland named Ernst Fehr, who'd recently published a study on altruism. I told the editor that Fehr was a potential competitor and I'd prefer not to share the article just yet, but off it went. Fehr is a very smart and very aggressive researcher, and I suspected he'd see very quickly that, after you'd measured oxytocin in the blood, the next logical step was to manipulate oxytocin in the brain.

Feeling the heat, I called Markus Heinrichs in Switzerland.

"I can't get approval here," I told him. "Let's collaborate on this. I'll work up the protocol, then we'll do the infusion in your lab in Zurich."

Heinrichs said, "Funny you should ask. Just two days ago an economist named Fehr called me up about the same thing."

The three of us got on the phone and we worked out a deal to collaborate at the University of Zurich, far beyond the purview of the Food and Drug Administration.

In Zurich we infused one hundred test subjects with Syntocinon, the heavily concentrated nasal spray that could deliver oxytocin directly to the brain. To create a control group for comparison, we infused a similar number with an inert solution. Neither of these substances causes any identifiable sensation, so nobody knew which one they'd just received.

We then had our hopped-up (or not) subjects play the Trust Game. Fortunately there were no heart attacks or incidents involving unwanted erections. On the distinctly plus side, the players who'd had oxytocin forced up their noses gave 17 percent more money to their partners than those who'd been infused with the placebo. Even more dramatically, half of the A-players on oxytocin were suddenly so trusting that they transferred *all* their money to their B-player, which was more than twice the number who were this trusting on placebo.

This nailed the clear demonstration of cause and effect I'd been looking for. Getting the brain to release oxytocin by way of a natural stimulus—the work I'd done at UCLA—was the real breakthrough. This was merely confirmation. But owing to peculiarities in the academic review process, the Zurich research was published first, and it caused a sensation. Next thing I knew I was in New York, explaining oxytocin to America on national TV, which has been a large part of my job ever since. There was one major benefit to all the publicity, however, and that was that I no longer had to go the indie route of begging and borrowing to support "the stupidest idea in the world." The John Templeton Foundation especially was very generous, providing $1.5 million for further research.

What this Templeton funding allowed me to pursue next was

the question of whether the oxytocin effect was limited to win-win situations like the Trust Game, or if it would increase generosity in zero-sum situations as well—those in which one person's gain is necessarily another person's loss. In the meantime, I'd found a way to finesse my FDA problems by making my own oxytocin inhalers with FDA approval. So I hadn't given up the indie approach to science altogether.

Using these handmade gizmos, I did another experiment back at UCLA in which I gave one participant $10, infused him with oxytocin, then asked him to give some of his money to another anonymous participant. That's all there was to it—just offer a gift. This wasn't a case in which the A-player being asked to donate has to think about how the B-player might react—instead it was a simple measure of altruism. How the other person would respond had no bearing on the outcome.

The result? Well, if you remember back to our Vampire Wedding, we showed that oxytocin was sensitive to the exact nature of human relationships. The Moral Molecule responds to human bonds, and in this simple test there was no bond to react to, or anticipate, or factor in when deciding what to do. So the oxytocin infusion did nothing.

But then we did the same infusion while having the subjects play what's called the Ultimatum Game. In this one there's a Proposer and a Responder, and their economic outcomes are joined at the hip. The Proposer gets $10 to start, but the catch is that in order to keep any of the money, he has to offer a share to the Responder. But the even catchier catch is that the Responder has to *agree to* the split or nobody gets anything.

In the Ultimatum Game, people almost always refuse a nine-for-me-and-one-for-you distribution just on principle. You might think $1—or even $2—is better than nothing, but apparently that doesn't cut much ice when the other person is being a greedy pig. When confronted with blatant unfairness, people seem to derive more pleasure (remember those reward areas of the brain?) from standing on principle than in making a buck. In the United States, and most other developed countries, in fact, offers of a split less than 30 percent of the pie are nearly always rejected.

When we did this study in which making money depended on social savvy (coming up with a win-win solution acceptable to both parties), an infusion of oxytocin caused generosity to surge by 80 percent!

So is the key to a better society getting everyone to simply snort oxytocin every few minutes? Aside from a number of practical obstacles, the infusion route isn't really necessary, especially since I've discovered that nature provides so many techniques for oxytocin release just in the course of everyday living. Dogs nuzzle and cats rub up against you to get their oxytocin high. Bottlenose dolphins pet and stroke one another. These feel-good behaviors all serve the purpose of strengthening bonds. In humans, the customs aren't really all that different.

## Touchy-Feely

I'd come into oxytocin research on the vector of trust. The next step was to see what I could learn about oxytocin and touch.

A few years ago, when I took my daughter to the first day of school, I was delighted to see that as children entered the classroom in single file, the teacher gave each of them a hug. When she got to the end of the line she found a six-foot-four dad (*c'est moi*) standing there saying, "Can grown-ups get hugs, too?"

I'd never been afraid to get physical, but after a decade spent studying oxytocin, I now warn everyone who comes to my lab that before our encounter is over, I'm going to give them a hug. It's amazing how just that statement of intent breaks the ice and allows people to connect in a more open way.

Unfortunately, the hug has gotten a bad rap. It may have been overexposure in the 1960s, but somehow pop culture has reduced it to a touchy-feely cliché, a tactile version of singing "Kumbaya." On the other hand, the Free Hugs movement became a sensation on YouTube in just the past few years. This oxytocin-inducing guerrilla effort began when an Australian using the pseudonym Juan Mann flew home from London and felt bereft that there was no one to greet him. Deciding that human warmth was what we all needed most, he made a cardboard sign saying FREE HUGS and went out to one of the busiest pedestrian intersections in Sydney. It took a while to break the ice, but eventually people started lining up. The movement went viral, with similar efforts in countries all over the world, millions of views online for Free Hugs videos, and the obligatory appearance for Juan Mann on Oprah.

Even more people may have been affected by a woman named Amma, called "the hugging saint," who is said to have embraced more than twenty-nine million devotees. On the spur of the moment I tried to see her once at a convention center in L.A. where she was offering her own distinctive approach to oxytocin release. I showed up late, left my car with a valet at the curb, and got in the door, but the hug line itself required an advance ticket, which I didn't have. Still, it was a powerful experience to see people overwhelmed by a simple, loving embrace. I watched for half an hour or so but then had to move on to an appointment. When I went back out to my car the valet shrugged and said, "No charge." I gave him a $20 tip.

From the perspective of my early research, one of the

interesting things about hugs is the role played by trust. Sure, hugs can be soothing and generosity-inspiring, but they can also be an unwelcome intrusion and a violation of someone's personal space. (Ever see the video of George W. Bush trying to give German chancellor Angela Merkel a spontaneous neck rub?) The difference is social context and social trust.

Hugs are a form of greeting, and greetings are always about establishing, or demonstrating, or rekindling, social bonds. Dogs sniff one another's rear ends. Humans shake hands, a custom said to have originated with men presenting the right arm in full view, to show that they weren't holding a weapon. But we also grasp the shoulder or elbow, and practice all sorts of variations on the kiss: one cheek, both cheeks, the left/right/left-again triple smooch. (There are, of course, further refinements of oral probing limited to romantic partners.)

Each of the gestures within a greeting is designed to convey certain information, much of which has to do with trust, and most of which we don't process consciously. But our unconscious usually does pretty well. Nothing feels quite as stiff as a misplaced hug (or as phony as an air kiss between "frenemies"). And nothing feels quite as comforting as a hug that transmits the full complement of warmth and trust that comes with the release of oxytocin.

In the lab I sometimes stroke a lab rat's belly when I want to release oxytocin and get the little beast to calm down. You can do the same thing with humans by rubbing your fingertips just between the ribs to stimulate the vagus nerve, which is rich in oxytocin receptors; this innervates the gut and causes people to relax and feel safe (my kids love this). But bringing human subjects into the lab to have their bellies stroked sounded problematic.

My next thought for increasing oxytocin was to bring people in just for hugs, but that, too, could get weird. Someone might go overboard, and then I'd get sued or frog-marched out of the university for instigating sexual harassment. We needed a form of touch that might stimulate the Moral Molecule but would allow all concerned to remain clinically detached. We needed someone in a white coat. We needed a licensed massage therapist! So I went to a school for therapists in Los Angeles and lined up three of their instructors to help us. That's how I spent $8,000 on massage without getting my back rubbed even once.

It was actually one of the hardest studies I've ever done. We would bring in eight to twelve test subjects at once, take their blood, give them a fifteen-minute massage, then have them play the Trust Game, then take their blood again. For our controls the routine was the same, except that we would simply have them rest quietly for fifteen minutes instead of getting the feel-good touch. They also came on different days from the massage group so they wouldn't feel bad about not getting the massage. But running back and forth trying to keep track of all these people in so many different postures of pleasure and repose made me feel like I'd wandered into an old episode of *Three's Company*.

Happily, the data we came up with were very much in line with what we'd seen when oxytocin release was prompted by trust. Across the board, those who got massages had a 9 percent increase in their oxytocin levels. But the real bonanza occurred with the B-players who'd received a massage and then received a transfer of money from an A that registered as a bond of trust. For this group (massage plus trust), the willingness to reciprocate by giving money back increased by 243 percent!

Warm physical contact (when welcomed and appropriate)

combined with a social bond proved to be the key to the kingdom when it came to prompting generous, pro-social behavior. This sounded like news we could use.

But as I said at the outset of this chapter, the most important through-line for my research is the movement from the center to the outer rings of that social "solar system" we first encountered at the Vampire Wedding.

We've seen how oxytocin facilitates contact between mates, and then bonding with offspring as well as with mates. With social mammals, the bond could expand to include a much larger kinship group, and even unrelated neighbors. But for social mammals burrowed in together, trust is high, physical contact an everyday occurrence, and the prospects for survival are held pretty much in common. So all the answers I'd found so far still begged the question: How can we generate and sustain this kind of family feeling among the far greater numbers of individuals that make up human societies, most of whom never see each other face-to-face?

I was pondering this issue a few years back, coming home on an airplane after being away from my wife and kids for nearly a week. I was pretty much burned out, so I shut down my laptop, took off my shoes, and watched the movie. Next thing I knew I was crying like a fool. The film I was watching was Clint Eastwood's *Million Dollar Baby,* the one about the female boxer who gets a head injury and doesn't want to live anymore. For some reason it really got to me, tears streaming, so much so that the flight attendant asked me, "Sir, are you all right?"

"Yeah," I said. "Thanks."

She smiled, and then I smiled. And then I said, "I think I've just found the next mechanism I want to investigate for stimulating oxytocin."

# CHAPTER 3

# Feeling Oxytocin

*The Circuit That Brings Us HOME*

The first year I was on the faculty at Claremont, my wife was doing her internship in Las Vegas, so once or twice a month I'd zip over to Sin City to spend some time with her. The apartment where she was staying had a pool, and while she was at the hospital I'd take my computer outside to enjoy the sun while doing some work.

One morning—it was a Tuesday, I think—no sooner had I set myself up at a table near the deep end than a mother came out with two rambunctious little boys. *Damn . . . there goes my peace and quiet,* I was thinking. *Why the hell does she have to take them swimming at ten in the morning?*

Sure enough, her five-year-old was a holy terror. He kept jumping up and down and shrieking, and she kept yelling at him to settle down and wait beside her while she put water wings on the two-year-old. But, being five, of course he immediately ran right around to the deep end of the pool and jumped in, sending a nice spray of chlorinated water up and onto my legs.

Then, also being five, he sank like a stone. I could see him thrashing around down there in about eight feet of water, but it wasn't what you'd call swimming. It was more what you'd call drowning. I glanced over at the mother and saw the look of crisis in her face. It was like *Sophie's Choice*. She couldn't leave the two-year-old in the water to go rescue the five-year-old, so what was she supposed to do?

Fortunately, she didn't have to do anything. I dove in and pulled the kid off the bottom of the pool. When I brought him over to her, coughing and crying, she was still so upset she couldn't speak to me. No word of thanks—she didn't even look up. She just kept yelling at the older boy as she grabbed both kids by their arms and wrangled them back to their apartment.

Almost every social animal has some kind of distress call. The interesting thing about humans is that we don't need to cry out in order to summon help. Often other humans can understand what we need merely by inference, helped along by the look on our face—sometimes just the look in our eyes.

I guessed that the kind of engagement that allowed this almost telepathic form of communication was oxytocin-based, but I wanted to understand more about how these kinds of messages were conveyed and what kind of mechanisms were involved. Mostly, I wanted to understand exactly what that surge of oxytocin *feels like*—the one that prompts the impulse for moral behavior.

One of my graduate students, Jorge Barraza, suggested a way to probe these questions using a five-minute fund-raising video produced by St. Jude Children's Research Hospital in Memphis, Tennessee. The first step was to edit it into two very different clips of one hundred seconds each.

In the A-version, you see a dad and a little boy having a nice day at the zoo. They hold hands as the toddler toddles along, they look at the giraffes, they laugh and talk. You might notice that the little boy has no hair, but other than that, it's all blue skies and happy childhood.

It's the B-version that hits you with all the heavy stuff. I've watched this thing a hundred times, and I still find myself tearing up when I show it in my lectures. It opens with BEN'S STORY spelled out in alphabet blocks. On the soundtrack there's music from a child's nursery, and then the camera pans from the blocks to a framed portrait of a cute little fellow with no hair. Then you hear the father's voice saying in a soft, Southern accent, "My son has a brain tumor . . ."

For one hundred seconds you share the father's pain in segment after segment, shot in hospital corridors and treatment rooms, with talk of chemotherapy and survival rates, and little Ben getting therapy to help him with his balance after four rounds of brain surgery. The most devastating part is when the dad looks right at the camera and tells you what it's like to know that his son is dying of cancer. To the strains of soft music he describes his relationship with his son, chokes back the tears, and then says, "You don't know what it feels like to know how little time you have left."

To explore the feelings this highly emotional footage might trigger, we brought together 145 volunteers, took their blood to establish a baseline, then divided them into two groups. One group watched the A-version of the film with the neutral emotional content. The other watched the B-version designed to bring tears to their eyes. Immediately afterward, we took everyone's blood again.

For those who watched the neutral version, there was actually a 20 percent *drop* in oxytocin. Note to aspiring screenwriters: Watching a father and son at the zoo for a minute and a half with no human drama can become, well . . . boring, and evidently these viewers became disengaged. But for those who watched the clip with all the heart-wrenching medical details, the spike in oxytocin was an incredible 47 percent over baseline. It made me wish I could have taken my blood before and after the experience at the pool when I thought the five-year-old might drown.

Obviously, we had found a very dramatic stimulus for oxytocin release. But what exactly was the force we'd unleashed? There were no wires or cables plugged in . . . no Wi-Fi . . . no skin being touched to connect one person's physiology to the others'. So what was going on to make the oxytocin surge?

To find out more about this mysterious action at a distance, we gave everyone who'd watched the videos a list of seven words that might characterize their experience, then let them rank these descriptive terms. Among the folks who'd seen the heart-wrenching version with the cancer details, two words were rated tops: *Distress* and *Empathy*. When we compared the change in blood values of each individual to the description each had selected for what the film had stirred up, the choice of *Distress* was directly correlated with an increase in the stress hormone cortisol. The choice of *Empathy* was directly correlated with a rise in oxytocin.

Interesting, but it still begs the question: How do we get from empathy to action?

When I saw the mother's face at the pool in Las Vegas, I knew what I had to do and I did it without thinking. But there was no risk to me—not even much of a cost. The situation was very different, however, in January 2007, when a construction worker named Wesley Autrey was in Manhattan with his two daughters. While they were waiting for the subway, a young man had a seizure and fell onto the tracks. With a train coming, Mr. Autrey left his daughters with a stranger, leapt down onto the railroad bed, and held the man down as the passenger cars rumbled by an inch or so over their heads.

Distress and empathy are not opposites—in fact they often run together. Moderate distress actually increases the release of oxytocin, which motivates us to engage. When reporters asked Autrey to explain why he did what he did, he said he leapt onto the tracks because he didn't want his daughters to see a man crushed and shredded by a subway.

Wesley Autrey became "the Subway Hero" all over the national media, an attribution he richly deserves, but in point of fact people risk their lives to save others this way quite often, and quite spontaneously. We know from evolution that for the helping impulse in us to be that strong—so strong that we risk life and limb— mutual assistance must be highly adaptive.

After having our test subjects watch the Ben's Story video, we didn't ask anyone to do anything heroic, but after the blood sampling and the search for the right word to describe their emotional experience, we did ask them to play the Ultimatum Game, the one in which a Proposer has to get the Responder to go along with a monetary split. Those who'd experienced the sharpest spike in oxytocin, who were also the ones who reported the strongest sense of empathy, made the most generous offers. These highly empathic viewers were also the most generous when offered a chance to donate some of their earnings to St. Jude's.

Just as I had been moved by watching Clint Eastwood's character in a film do an incredibly generous thing for a stranger he'd become attached to, our brains don't differentiate between a person in need in a flickering image and a person in need in front of us. This is why we can be moved by great movies, great music, and great art. Through oxytocin release, these products of the human imagination connect us to the entire human family. This is what we want most as social creatures.

## Where We Decide to Do the Right Thing

*Empathy* is a word more often associated with greeting cards than with science, but it isn't just a feel-good emotion sitting in some special compartment shaped like a Valentine's Day heart. Although some areas, such as the gut, are more densely lined with oxytocin and stress receptors, there is no one, isolated compartment for the emotions to fit in, just as there is no cartoon frame in our head called "the mind" where a bulb lights up when we get a good idea. These two dimensions of our experience—the emotions and the intellect—while often in conflict, and sometimes working at cross purposes, are actually two parts of the same whole, and that whole is the body. We have thoughts and we have emotions, and both are the output of physical systems that have been evolving in animal cells and tissues for about three billion years.

William James, the father of experimental psychology, defined emotions as the physiological changes your body undergoes when your senses pick up certain signals from the environment. Nipple stimulation, for instance, releases oxytocin in breast tissue, and that not only causes a mother's milk to flow but also changes her emotional state. Because of oxytocin, her focus draws in on her immediate surroundings, her anxiety level drops, and her brain releases dopamine and serotonin to give her pleasure. These emotional changes are vital to increasing her willingness to tolerate this pesky creature (or maybe a whole litter of similar creatures) demanding her attention, as well as her limited metabolic resources.

These changes in the internal state of the organism take place almost instantly, without deliberate control, and without conscious awareness. When humans become aware of these kinds of physical sensations, so much so that we can identify them and give them labels like distress or empathy (or fear or happiness or contentment), that's what James called a "feeling." A feeling is our conscious awareness of the emotion. The emotion is the physical experience that's happening down in our cells and tissues.

But once again, how does an emotional change—an experience that touches us—actually do the "touching," especially when the contact isn't skin-to-skin? How can watching a movie (or seeing a child about to drown or left crying on the street, or seeing your grandparents holding hands) cause the kind of chemical changes that alter your outlook as well as your behavior? How can hearing a single word make testosterone surge to the point that men become raging maniacs in bar fights? How can seeing a certain smile across a room create the full-body tingle we call romantic love? Then again, how did being trusted in our early oxytocin experiments change participants' emotional state to the extent that they wanted to reciprocate with the stranger who'd trusted them?

We readily accept the physical nature of the emotion we call fear, which doesn't require physical contact, because of the way in which it grips us immediately. When you hear footsteps behind you in a darkened parking garage late at night, the sound comes in through the auditory system and registers in the part of your brain that responds to threats—the amygdala—which triggers the fight-or-flight response via the release of stress hormones. These make your heart begin to thump and your palms begin to sweat— two of the dramatic physiological changes that *are* the emotion of fear. When we become aware that all this is happening—that's the *feeling* of being afraid.

The tricky thing about empathy is that the whole effect is

much more subtle. It also requires that several other factors coincide.

I've already mentioned the release of oxytocin in breast tissue, which triggers the flow of milk as well as a certain warm and loving experience. Even that primal, maternal reaction can happen at a distance, without touch or wires or cables. Oxytocin can make a mother's milk flow and her outlook turn all warm and fuzzy whenever she sees or smells a baby, or simply hears one cry. But the emotional response is based on a kind of cellular memory laid down by oxytocin ("that's what my baby smells like"). Animals bred without the ability to produce oxytocin have permanent social amnesia.

Empathy in humans requires that same kind of cellular association. Sights and sounds of trust, or distress, or compassion can trigger memories that take us back to our earliest experiences in relationship to others. These memories trigger the release of oxytocin, which ultimately creates the sensations at the level of cells and chemicals and brain structures that we identify as empathy.

But then what? Do we just get all warm and fuzzy and then rationally *decide* that we will now behave more altruistically?

About two hundred years ago German philosophers began talking about "feeling into" works of art and architecture. Eventually, this feeling-into concept made its way from the discussion of paintings and buildings and into the realm of psychology. The term they used was *Einfuhlung,* which made its way into English as *empathy*—a new word based on the Greek *pathos,* combined with the Greek prefix meaning "in." Theodor Lipps, a nineteenth-century philosopher and psychologist, explained empathy this way: "When I observe a circus performer on a hanging wire," he wrote, "I feel I am inside him." But Lipps's explanation was pretty much just restating what our old friend Adam Smith had expressed in *The Theory of Moral Sentiments* in 1759: "When we see a stroke aimed and just ready to fall upon the leg or arm of another person, we naturally shrink and draw back our own leg or our own arm, and when it does fall we feel it." Smith had argued that imagining "our brother upon the rack" was enough to make us "enter as it were into his body, and become in some measure the same person with him."

All of which sounds like what happens when you see someone looking sad or worried and you get the physical sensation of sadness or worry yourself, or when you hear someone laugh and you can't help laughing, or at least cracking a smile.

Freud found this empathy concept hugely significant, and Heinz Kohut and Carl Rogers went on to make it a central feature of twentieth-century psychotherapy. But then the developmental psychologist Jean Piaget gave it a twist, emphasizing the degree to which knowing other minds, even to the point of establishing empathy, requires intellectual perspective-taking. After all, it would be hard to say that you're being empathic when you're shaking uncontrollably and shrieking in terror after witnessing some tragedy. Empathy is calmer and more measured, more about the other person than about yourself. In short, it's the difference between a surge of oxytocin and a surge of adrenaline.

The debate over the essence of empathy persisted for a long time, and then neuroscience came along, bringing with it the brain scans and blood tests that allow us to look for answers down where the action is.

Jean Decety, a neuroscientist at the University of Chicago, helped decipher one aspect of empathy with a number of studies that relied on our perceptions of pain. He showed pairs of photographs to test subjects at the same time that he scanned their brains using functional magnetic resonance imaging (fMRI). One photo would be of something ordinary—a hand on a pair of pruning shears cutting a branch; the other would be the same hand and shears, only this time the hand would be squeezed between the blades. One picture would show a bare foot beside an opening door; the other would show the opening door lined up to dig into the top of the foot. When the pictures switched from ordinary to gruesome, specific areas in the subject's brain would light up. These specific areas were the same ones responsible for coordinating emotional responses to your own pain. As far as your brain's response, then, any pain you observed in the photographs was pain that was happening to *you*.

Giacomo Rizzolatti at the University of Parma took the investigation deeper still by putting electrodes into monkey brains. Whenever a monkey reached for something—usually a peanut—neurons in the pre-motor cortex would fire. But when one of the researchers reached for a peanut while the monkey watched, these same monkey neurons fired. It was as if the animal had picked up the peanut himself. If the experimenter put a nut in his mouth, the same monkey neurons fired that fired whenever the monkey put a nut in its mouth. These "mirror" neurons would fire even when the critical point of the action—the person's hand actually grasping the nut—was hidden from view. Even hearing the action of the shell being cracked was enough to trigger the response. With a minimal amount of information, the monkey brain could go on to imagine the rest.

Rizzolatti wanted to see if he could demonstrate the same kind of effect in people, but scientists aren't allowed to mess around quite as freely in human brains. So he did the next best thing. Working with neuroscientist Luciano Fadiga, he examined the twitching of hand muscles—a signal that the hand is about to move—while human test subjects watched an experimenter grasp various objects. The hand twitching was the same when the participants observed an action as when the participants themselves actually did the grasping. And it didn't matter whether or not they could actually see the experimenter's hand moving through the entire sequence. The brain created a narrative that filled in the gaps.

For certain kinds of information, then, the brain simply breaks down the barrier between ourselves and those around us, so much so that we want to treat them as well as we'd treat ourselves. Which is an idea that sounds oddly familiar. In fact, it reminds me of some of those ancient moral traditions that were part of my early education. Empathy, in effect, creates a physiological version of the Golden Rule. This means that when a situation we see or learn about causes us to "do unto others as we would have them do unto us," it is in part because we are *literally* experiencing another person's pleasure or pain as if it *were* our own.

It turns out that Adam Smith was right when he said that "fellow feeling" was the basis for moral action. Two hundred and fifty years after *The Theory of Moral Sentiments*, we're able to offer a detailed explanation of the process that Smith could only imagine. We can trace empathy from the initial surge in oxytocin, to the release of dopamine and serotonin that makes the experience both pleasurable and something you want to repeat, to the social engagement that emerges as a result. The neuroscience explains not only the Golden Rule but also the Confucian concept of ren (benevolence) and the Buddhist concepts of metta (loving-kindness) and karuna (compassion).

But there's still a pretty big difference between having neurons fire—which can be simple stimulus and response—and experiencing (and then acting more virtuously as a result of) what we call empathy. A mother rat will respond to stress by clustering her young beneath her—stimulus and response. And almost any rat in a lab experiment will stop pressing a bar to obtain food if it detects that the bar delivers an electric shock to another rat nearby. But that's not because rats are empathic. They're simply smart enough to suspect that what's bad for the rat next door might very soon become bad for them. Even in species as intelligent as monkeys, while a mother will actively protect her young, she doesn't provide anything that looks like empathic cuddling and soothing, even when her offspring has been bitten. For sure they don't get into ren, or metta, or karuna.

Jean Decety identified four elements he considered essential for empathy, but these aren't arbitrary qualifiers. Instead, each represents one of four distinct processes, each taking place in four different areas of the brain in response to seeing Decety's photographs.

The first of these is shared affect, which pretty well sums up the kind of mirroring and co-experiencing I've just described.

The second is awareness of the other as apart from the self, which is a cognitive capacity called Theory of Mind. This begins to emerge in humans at about age two. At the same time that we begin to recognize ourselves in mirrors, we learn that Mommy is not just an extension of us, and we begin to understand that other people have thoughts and feelings that are independent of our own.

The third is the mental flexibility to nonetheless put ourselves "in the other person's shoes."

And the fourth element is the emotional self-regulation necessary to produce an appropriate response, which relies on a special ability located in the pre-frontal cortex called executive function.

This is what allows us to avoid screaming every time someone makes us angry, or crying every time we see a sad child. Executive function is what trauma surgeons and first responders need in large doses in order to remain compassionate in the face of hideous situations, while also remaining sufficiently detached to do what needs to be done to help.

## The HOME Circuit

Because Decety's work relied on pain perception, he omitted the most vitally important element that triggers the entire effect: oxytocin-producing neurons and oxytocin receptors. Oxytocin, combined with the two feel-good neurochemicals it releases—serotonin and dopamine—activates the Human Oxytocin Mediated Empathy (HOME) circuit. Dopamine reinforces the smile of thanks we get when we treat others well, and serotonin gives us a mood lift. It is this HOME circuit that keeps us coming back and behaving morally—at least most of the time. As we'll see later, stress, testosterone, trauma, genetic anomalies, even mental

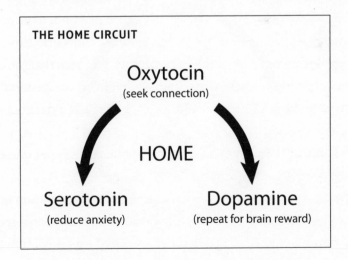

THE HOME CIRCUIT

Oxytocin
(seek connection)

HOME

Serotonin
(reduce anxiety)

Dopamine
(repeat for brain reward)

conditioning can inhibit these effects. But as long as we keep these influences from taking over, the system is self-reinforcing.

Because of the range of influences we're subject to, human beings can be both good and bad, but in stable and safe circumstances oxytocin makes us mostly good. Oxytocin generates the empathy that drives moral behavior, which inspires trust, which causes the release of more oxytocin, which creates more empathy. This is the behavioral feedback loop we call the virtuous cycle.

Observing another person's distress catches our attention, and we experience some of what they're experiencing. This can cause oxytocin release, but not if our own distress is above a certain threshold. Nature assumes that if we're in dire straits ourselves, we can't so easily afford to invest time and resources in helping another. High stress blocks oxytocin release—in most cases, oxytocin is doubly inappropriate for someone who's being pushed to the edge of survival himself. Oxytocin not only drives empathic

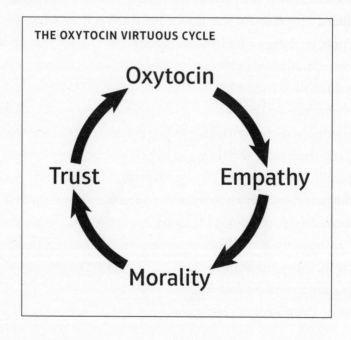

**THE OXYTOCIN VIRTUOUS CYCLE**

Oxytocin

Empathy

Morality

Trust

concern (compassion)—which might get in the way of fighting for your life—but also damps down the amygdala, the brain structure where anxiety is registered and regulated.

Choosing to be altruistic, or even heroic, in overriding our self-protective anxiety is another matter, and whether we will sacrifice to help another often depends on our degree of closeness. We would run into the burning building without a second thought to save our own child; soldiers sacrifice themselves to save the buddies in their unit who have become their brothers in arms. Although people still do it, we're less likely to take a risk to help a stranger, and being in a fear-provoking place is not the only obstacle. The likelihood of altruistic action can decrease depending simply on how much we're caught up in our own problems at the moment.

If the Subway Hero's daughters had been so young that he had to worry about them toddling onto the tracks, he might not have jumped to the young man's aid. If he'd been caught up in some intense discussion with one of his children, that, too, might have precluded his spontaneous action. If, instead of working in construction, he'd been a high-status investment banker who'd somehow wandered into the subway, his feeling of social distance from the man in need might have gotten in the way.

It's because of their ability to set aside all their self-interested or self-involved concerns, fully experience an empathic connection, and then risk everything on behalf of someone else that we call people like Wesley Autrey, the Subway Hero, heroes.

But nature has also worked a bit of moral judgment into these oxytocin-driven, physiological reactions. We readily help kids and cute animals, in part because we know that whatever trouble they're in, they can't really be held accountable. We're less likely to be so understanding and forgiving when it comes to homeless adults or drug addicts. For some people, teenage girls who get pregnant deserve the same cold shoulder. "You've made your bed," they say, "now lie in it."

This tendency to judge rather than help is partly the result of a spot in the prefrontal cortex called the subgenual cortex. It's full of oxytocin receptors, and it appears to modulate the degree of empathy by regulating the release of dopamine in the HOME circuit. No dopamine means no reward from engaging with the other person, which makes it less likely that we'll reach out empathically.

So here again, oxytocin maintains the balance between self and other, trust and distrust, approach and withdrawal. When the brain releases oxytocin, the balance shifts toward empathy, and we contribute resources to help others. When the oxytocin surge fades, we move on from the feeling of empathy, the HOME system resets, and we're ready to evaluate the next interaction that comes along. When testosterone and other pro-punishment factors take over, we're ready to throw stones rather than a lifeline.

But the question remains: Why has natural selection steered us toward compassionate behavior, which, at least in the short term, looks very much like a disadvantage? After all, don't nice guys finish last?

Well, even before animals did much in the way of warm and fuzzy attachment, there was serious, competitive advantage in knowing as much as possible about the internal state of other creatures. When another animal is out to get you, a warning that he's not in a charitable mood can help keep you alive. By the same token, knowing that the other guy is content could save you a lot of energy, as well as unnecessary scar tissue.

The most primitive source for this kind of information was the same signaling system we saw with the amorous lobsters—chemical detection. We still have a vestige of that messaging system today, lodged, appropriately enough, in the most evolutionarily ancient part of our brain. That vestigial system is called smell.

But well before we came along, signaling and detection needed to become more subtle and discriminating than simply relying on smell, because "other creatures" had come to mean far more than a threat, a meal, or a mate. Mammal babies, for instance, depend on their mothers for care, so the more Mom knows about what's going on inside her little bundle of joy, the more effective she's going to be in keeping said bundle alive. Is Junior fearful and in need of calming? Is the Little Princess hungry and needs to be fed? It's this primordial mothering role—mother "love," if you will— that created the more granular, sensory perceptions that eventually linked oxytocin with empathy. (It also helps explain why females have freer access to both than males. In every experiment I've designed for humans, women release more oxytocin than men.)

Because baby mammals depend on the breast for nutrition, there has to be a strong bond between mother and child or the child will die. Mom also has to be tolerant and attentive for long periods of time, and Junior can't be such a loner that he's prone to wandering off. So this is where bonding became a necessity, not a feel-good frill. As biological bonds became more sophisticated, they moved beyond chemical imprinting to include all sorts of other sights and sounds and complex associations, and after millions of years they reached the level we now call emotional attachment.

For the highly intelligent and highly social mammals known as apes, there's long been more to early survival than just sucking on a nipple and staying close. Becoming a proper primate requires social learning, and that process needs to start as soon as the ape in training can focus its eyes.

Newborns—and this is true of both chimps and humans—begin to focus on faces and mimic facial expressions only a few hours after birth. Open your mouth and they'll open their mouth. Stick out your tongue and they'll do same. They're trying on these social gestures, working to master them and weave them into their neural wiring. When that wiring is done right, it solidifies the most fundamental bonds, which help us through the early years but also set us up to do well with the emotional demands and pleasures of later life.

For the human brain—except when autism is a factor—the human face is the most significant object in the universe, and it attracts our attention like nothing else. The fascination begins the moment we're born and continues throughout life. Even passersby in grocery stores get drawn in to oohing and ahhing at babies because baby faces are so "cute." This is no accident: Natural selection sculpted big round eyes and chubby cheeks to ensure maximal appeal, in order to maximize the chances of survival. The cartoon cuteness of babies is called neoteny, and robotics engineers work it into their designs when they want people to easily relate to their computerized creations.

At six weeks some human babies can remember and imitate a gesture performed by an adult on the previous day, which helps the baby identify the individuals who really matter, such as Mom, Dad, Grandmother. So none of this "peek-a-boo" business is just an idle pastime. It's all about tuning up for social life, which for hominid apes is the only kind of life there is.

After two or three months, both chimp and human babies give up being face freaks. They've made the most basic neural connections they needed to make, and it's time for social learning to take over, which will lead to a wider array of bonds. Human babies decide very quickly whom they will trust (Mom, Dad,

Grandmother, their favorite babysitter) and whom they want to remain wary of (just about anybody else who doesn't come with the implied approval of those immediate caregivers). But over time, social savvy needs to move beyond such a limited horizon.

## Social Networks from the Ground Up

During the millions of years of our development as social mammals, our individual survival depended on how well we fit in with the group, and group survival depended on how well each member cooperated. By adulthood, our hunter-gatherer ancestors would sink or swim based on their ability to judge who was lying and who was telling the truth. Whom could you count on to watch your child? How could you work out a better deal when it was time to overthrow the leader and form a new coalition? And whom should you be willing to sacrifice for when you sensed that they needed help?

Human infants immediately orient to the distress of others and respond to crying with their own cries of distress. Beginning at about eighteen months, a human will almost always offer assistance to another baby, or to an adult, if it's possible to do so. For chimps, even in adulthood, the likelihood of one animal helping another is about 50–50. For chimps, helpfulness depends on kinship, familiarity, recent interactions, and whether or not the animal in question can simply pay attention long enough to be helpful at that particular moment.

Human babies not only are predisposed to help others but also show a preference for others who play nice, and an aversion to those who don't, even when the "players" are inanimate objects. This has shown up again and again in studies in which infants view a film about geometric shapes, made by psychologists Fritz

Heider and Mary-Ann Simmel in 1944. In this animation there's a box with a door, a ball, a small triangle, and a big triangle that appears to menace the smaller triangle and the ball. As indicated by their gaze, babies are drawn to the "nice" geometric shapes, and they try to avoid the "mean" one. A computer scanning this film, like many people with autism, would see nothing more than shapes moving in a cartoon landscape. The socially adept human brain, however, with its propensity for meaning-making, sees a drama unfolding with good and evil, villains and victims.

Throughout our long evolutionary history, human infants survived at a higher rate when there were two committed adults looking out for them, which put a premium on what's called the pair bond. As every schoolchild knows, this kind of bond between men and women is facilitated and sustained by loving touch, including sex, as well as by empathic concern. Both touch (sex) and empathy involve oxytocin, and both are deeply ingrained in our concept of morality.

Similarly, cooperation within the larger group was enhanced by the oxytocin-driven empathy and trust that round out the virtuous cycle. These biological impulses—which say, in effect, *Stay calm and cooperate*—came to be reinforced by social behaviors such as gift-giving and other ritualized exchanges that were then passed along as part of culture.

Among apes, the primary stay-calm-and-cooperate ritual is grooming, which involves picking around in one another's fur, but it's not just about collecting ticks and mites. Fingers combing through fur releases oxytocin to calm the nerves, lower the heart rate, and allow blood pressure to drop down into the low range of normal. Apes spend roughly 10 percent of their time massaging one another this way—because getting everyone to stay calm and cooperate is vitally important to survival.

Grooming is also the easiest way to bestow a favor, and even among apes, the social brain is well equipped for keeping score on who's a taker and who's a reciprocator. Studies show that animals that groom each other more in the morning are more likely to share food with each other in the afternoon.

A more extreme stay-calm-and-cooperate custom developed among the chimp's cousins (and ours) the bonobos, which run their troops like a hippie commune, using sex to mellow out any bad vibes that come along. A typical female greeting is to have oral sex. Meanwhile, males hang from tree branches rubbing their penises together like crossed swords. And bonobo youngsters go at it like there's no tomorrow, acting out their own version of "Bonobo see, bonobo do." All of which keeps the individuals in the troop flooded with oxytocin, which means that bonobo society is about as peaceful and cooperative as a society can get. Only trouble is, bonobos haven't made a hell of a lot of progress in the past seven million years.

Then again, neither have the more aggressive and competitive chimps. Moving on out of the rain forest was left to their cousins—us—who found the sweet spot that combined healthy competition with a high degree of cooperation, the push and glide of testosterone and oxytocin. By triggering the release of dopamine and serotonin, oxytocin created the motivational pathway I've called HOME. Don't push too hard, don't get too upset, give as well as take. That's where the virtuous cycle gets its virtue.

Among family and close friends, we humans hug when greeting or parting, and we've been known to give a back rub when someone we care about is upset. But the stay-calm-and-cooperate social behavior that most closely resembles the apes' propensity for grooming is conversation. Anthropologists who study primitive societies are often struck by how much time these people spend simply lying around swapping stories, and how often the stories are about who is sleeping with whom. Sure, there are the myths and legends that need to be passed along to keep a culture alive, but juicy details about one's neighbors seems to be the preferred topic. Yet this idle chatter isn't just idle chatter. Conversation—especially conversation rich in social content—builds trust, which has the effect of a verbal massage or aural grooming and releases oxytocin. It also provides critical information bearing on the life of the group. What are the new alliances taking shape? Which guy can be trusted to stick around, and who's going to be a heartbreaker (and offspring abandoner)? Today the same principle applies to questions like who's a great mechanic, and which guy down the road is going to sell you a new starter motor you really don't need?

The gossip habit is so deeply ingrained in humans that now, in the media age, we've built huge industries around the exchange of trivia about characters on trashy reality shows as well as news about the latest Hollywood rehab or divorce. And where are the best places for sharing this kind of information, which so often leads to the sharing of one's own personal secrets? Where trust and physical contact (and grooming) create an oxytocin-rich environment: the hair salon, the barbershop, the locker room, the yoga studio.

For men, the verbal exchange is less often about movie stars and more often about sports figures, but the premise remains the same. All the stats and detailed descriptions being batted back and forth serve no purpose—except for the vitally important purpose of easing stress while strengthening human bonds.

Humans also build oxytocin-rich bonds through participation in sports and other friendly games, just like the dogs you see nipping playfully at one another down at the park. And the same stay-calm-and-cooperate rules apply, even when there is no referee to enforce them. If a six-foot-eight college hoops star goes at it too hard, turning a friendly pickup game into a personal highlight reel, the other guys will head to the showers. The unspoken rules of the friendly game incorporate a high degree of trust. You play clean, and you call fouls honestly, and you don't keep dunking over the sixteen-year-old who's not that good and stands a foot shorter than you. The trust and the physicality, combined with the moderate stress of friendly competition, help to create strong friendships among guys who may play together for years, and chitchat together every week in the locker room, and never even know one another's last names. (Oxytocin release also accounts for all that butt slapping that would get you a punch in the nose at the office but is perfectly okay during the game.)

We also further human bonds by continuing to mirror and mimic people throughout life. As infants we start out obsessed with faces, but our brain's focus on other people never goes away. If you and I stand facing each other and I cross my arms, you're more likely to cross your arms. If you rub your nose, I'm more likely to rub mine. We adopt the speech patterns of others, and any gesture, from laughter to yawning, can be contagious. People mimic the mannerisms of complete strangers, even when it's highly unlikely that there will be any future relationship or rapport. These kinds of reactions are not only involuntary but so fast that our imitations occur prior to our awareness of them.

Therapists know that clients often rate the interaction more highly when the counselor has mimicked the client's postures. Classes in which observers noted a high degree of physical mimicry are the same classes that the students themselves rated as being high in rapport. People who've been mimicked—even when they haven't consciously noticed—later reported having a more favorable impression of the person doing the imitation. So when there's a desire to affiliate—you're with the boss, or the local hero, or a love interest—the degree of behavioral mimicry increases.

Sometimes our tendency to mimic others creates empathy where it might not belong. In one experiment, participants who were instructed to consciously *resist* mimicking their partners were much better at detecting liars. (Which suggests that if somebody had warned me not to mimic that guy at the ARCO station in Santa Barbara, it might have saved me a hundred bucks.)

At a horse race, we lean into the turn with the rider we're rooting for. Watching a softball game, we crane our necks along with the center fielder as she stretches to make the catch. But good athletes are even more attuned to one another than their fans are to them, anticipating teammates' moves. Rapport contributes to synchronization, and synchronization contributes to rapport, which can make all the difference in beating a full-court press or making a double play, landing a jumbo jet with engine trouble, doing thoracic surgery, or serving up 126 perfect dinners in the kitchen of a busy restaurant.

It's at this point that the cognitive can join forces with the emotional to create the Holy Grail of every coach or CEO—having everybody thinking the same thing the same way, focused on the same goals. Psychologists call this co-cognition, the ability to know instantly what someone else's gesture or movement is all about, what its goal is, and how it relates to other actions and events, past, present, or future. Think of the "no look" pass in basketball, or the way jazz musicians intuitively play off one another when improvising. There's nobody barking orders, but everyone knows what to do.

Early humans, even after they had developed speech, still needed this kind of wordless synchronization to bring down big game or to corner and trap small game. Hunter-gatherer women benefited to the extent that they shared an almost collective consciousness for the tasks at hand. Is every child accounted for? How widely can we spread ourselves out without inviting an attack from a predator? This ability allowed our ancestors to make quick and sometimes vitally useful inferences, relying on physical cues and sensations that are, themselves, imperceptible to our conscious minds. And it all begins with the bonds created by oxytocin.

The virtuous cycle, with oxytocin front and center, is still the glue that holds society together. But as we've made clear all along, oxytocin doesn't have free rein. Other factors compete to influence us, and one of these is just as deeply ingrained in our sexual origins as oxytocin.

# CHAPTER 4

# Bad Boys

*The Complications of Gender*

I was twelve thousand feet above the desert in a hollowed-out prop plane, getting badgered by a grad student wearing a parachute over his lab coat. He kept asking me to add up pairs of numbers, but I couldn't concentrate because people were walking to the rear of the plane and disappearing. I found this especially distracting because I was sitting in the lap of a six-foot-three jump instructor to whom I was harnessed more tightly than I thought advisable, and in just a minute or so he and I were going to duck-walk together toward that same rear hatch and step out into the sky.

I have a serious fear of heights, but in anticipation of my first skydive—anything for science—I'd been dosing myself with a testosterone supplement for a week. The night before the jump I took my blood to establish a baseline for testosterone, oxytocin, and cortisol. Immediately after landing I was going to jab a needle in my arm to take blood again to measure the effects of a seven-thousand-foot free fall at 120 miles an hour. Whether it was the artificially ramped-up male hormone, or just plain ol' fear and excitement, by the time my instructor and I did a backflip out the rear of that plane I yelled "Geronimo!" just like those guys in the old war movies.

Testosterone will make people do some very strange things. Truth be told—the strange-acting people tend to be men, not women. It's testosterone that prompts male risk-taking, male violence, as well as the gender's most characteristic behavior: the reckless pursuit of sex, regardless of consequences.

There have been, in fact, so many newsworthy males brought low by their libidos in the past few years that it's hard to keep up. The prize for absolute number of simultaneous affairs goes to golf legend Tiger Woods. For worst name associated with a sex scandal, the winner is Anthony Weiner. For shameless audacity, the standout may be former South Carolina governor Mark Sanford, who said he was "hiking the Appalachian Trail" when he was actually below the equator with his Argentine mistress. (Then again, Arnold Schwarzenegger's having a child with his housekeeper, and keeping it a secret from his wife for ten years—all the while continuing to employ the housekeeper—was, arguably, lower still.) For political impact, there's Italian prime minister Silvio Berlusconi's preoccupation with teenage girls, which led to a revolt in Rome, and of course Bill Clinton's fling with an intern that nearly brought down his presidency.

But it's not just alpha males who get caught with their zippers down. Recently the world was drawn to the plight of thirty-three Chilean miners trapped for two months by a cave-in twenty-three hundred feet below the surface. As the ordeal wore on, drama became soap opera as an overabundance of loved ones gathered at the site. This led to the awkwardness of beleaguered wives discovering beleaguered girlfriends holding vigils for (and claiming workers' benefits from) their trapped husbands/boyfriends.

As a neuroscientist, I know that males have slightly less integration among the various parts of the brain than females, which makes it easier for men to compartmentalize the emotional and the erotic as separate categories—and activities.

But I also know that the real driver of the Bad Boy bus is the oxytocin antagonist known as testosterone. This hormone is present in women as well as men—it's just that men have ten times as much. Testosterone is great for athletic performance because it increases muscle mass and bone density, which is why sports heroes have been known to cheat by injecting synthetic testosterone precursors in the form of anabolic steroids. It's also useful when you need to run into a burning building to rescue people or to hit the Normandy beachhead under machine-gun fire, or in any other situation that requires risk-taking, physical courage, strength, and speed.

But the fact is, testosterone also causes a hell of a lot of trouble, and not just in the area of intimate relationships. Most crimes are committed by young men, and most murderers are males in the age range of twenty to twenty-five. (Murders by women are so rare that they don't even show up meaningfully in the crime statistics.) Young men have levels of testosterone that are twice the levels of older men, so the term *testosterone poisoning* for this age group is no joke.

Given all I've said about the role of oxytocin in sustaining social cooperation, and what I've said about the role of social cooperation in human survival, you might wonder how the molecule of reckless and generally anti-social behavior, testosterone, ever made it into the twenty-first century.

Well, a lot of testosterone did get voted off the island. In prehistory, as in the Wild West, males ramped up with too much of the stuff were often culled from the gene pool fairly early. They took absurd risks that got them killed, or they were so disagreeable and disruptive that they failed in the mating game, or the tribe (or the townspeople) simply unfriended them by bashing them in the head (or gunning them down with a six-shooter).

But even though pro-social behavior was the distinctive trait that allowed *Homo sapiens* to outcompete fiercer—and more fiercely competitive—animals like our cousins the chimp, we still needed testosterone in our environment of evolutionary adaptation. There were predators that needed fending off, potential sources of protein that sometimes fought back, and large objects like rocks and logs that from time to time might need to be moved. As a result, the physical strength, endurance, and aggression supplied by testosterone was essential to staying alive long enough to reproduce.

Moreover, in the contest for survival of the fittest, every group of early humans or pre-humans was at risk not only from big, fierce animals but also from big, fierce neighbors who competed with them for resources, including the food they needed for their children. To stay in the game, every tribe or troop needed some big, fierce players on their own team, even if these guys were sometimes lacking in the sensitivity (and sexual fidelity) department.

But the primary reason testosterone—and males—came to exist in the first place was to improve the quality of the gene pool by competing for the opportunity to mate. Over time that same testosterone-driven struggle to get one's own genes into the next generation created the drive for social status, which fueled the drive for better ways of doing things. Neither of which necessarily dovetailed with being absolutely the nicest guy in the world. So even today, testosterone is still around to increase motivation and drive—and not just the sex drive—in all humans, female as well as male.

Over millions of years of evolution, then, what emerged was a two-pronged approach to keeping the species alive. Either gender was capable of violence and competition and aggression, as well as bonding and compassion, but men (high in testosterone) were hormonally predisposed to take the lead on the former, while women (releasing high levels of oxytocin in response to stimuli) were hormonally predisposed to take the lead on the latter.

Women have been known to show up on the police blotter, cheat on their husbands, and commit fraud, embezzlement, and child abuse, but the fact is that they are, on average, more empathic, more other-focused, more trustworthy, more generous, and more charitable than men. In our Trust Game, while the average amount returned by a male B-player was 25 percent, the average returned by females was 42 percent. On the negative side of the behavioral ledger, 30 percent of men returned less than 10 percent, but only 13 percent of women were that coldhearted. When it came to being a real stinker, 24 percent of men returned absolutely nothing—which was true of only 7 percent of women.

But when we look at the really big picture, a curious fact stands out: The male licentiousness that produced our rogue's gallery of high-profile philanderers appears bound up with a completely incongruous male desire, which is to punish offenders. Despite my gender's well-documented moral failings, we're the ones who elect ourselves the enforcers—the hanging judges, the preachers who condemn the sinners, the hard-ass drill sergeants, the no-excuses CEOs.

There's no better example of this contradiction than Eliot Spitzer, formerly a talking head on CNN. As New York State attorney general, this married father of three daughters was known as a relentless crusader against graft, corruption, and all manner of wickedness. As governor of New York, he became best known as

"Client #9," the devoted patron of a high-end, and highly illegal, Washington, D.C., escort service.

Back in the 1990s the five top congressional leaders who drove Bill Clinton's impeachment seemed to have been cut from the same testosterone-soaked cloth. For months these men excoriated the president for his Map Room cigar episode with Monica Lewinsky. But before the dust settled, each of these family-values Republican politicians had been exposed for conducting his own extramarital trysts, at least one of which had produced an illegitimate child.

So does testosterone simply make men hypocrites as well as dirty dogs?

## Rub It Only on Your Shoulders!

To explore the role of testosterone as the Moral Molecule's evil twin, I needed to be able to introduce it into the behavioral equation in a precisely regulated way, much as we'd done in our with-and-without comparisons that began by infusing participants with oxytocin. Luckily, in this case, there was a medication for ramping up testosterone readily available, and it was FDA-approved.

AndroGel is the synthetic testosterone preparation I took before my skydive. It's a gel that's absorbed through the skin, and it comes in convenient, single-serving foil pouches, sort of like antibacterial hand gel. In fact it looks and smells like hand gel—which made it pretty easy to come up with a placebo for the "without extra testosterone" condition in our studies.

Anytime I'm going to subject someone to an experience in the lab, I always try it myself first, just so I'll understand what I'm asking someone else to put up with. So I got a prescription for AndroGel and for two weeks rubbed the stuff into my shoulders at the same time every day. The peak effect comes sixteen hours later. So the next morning, and every morning for the two weeks thereafter, I woke up feeling like I was nineteen again. I had monumental workouts at the gym. I didn't need nearly as much sleep, and I walked around with the cocky self-confidence (and libido) of the high school football player I'd once been.

Luckily, being a testosterone-crazed alpha did not seem to affect my ability to relate closely with my kids, or with anyone else for that matter. And I'm happy to report that I emerged from the experience without getting into any fights, or even yelling at anyone over a parking space.

Our participants for the AndroGel experiments reported for duty in the afternoon. Not too surprisingly, our recruitment ad, which had asked for young males to experiment with testosterone, pulled in a roomful of athletic, weight-lifter types. We had excluded women from the study because of the risk that testosterone infusion could create reproductive problems for them.

We took four tubes of blood from these men to measure their baseline testosterone levels, and then we monitored them as they applied the gel. "Rub it only on your shoulders, gentlemen . . . not down there." They came back first thing the next morning to have their blood drawn again, and to play the usual economic games. We then had them come back six weeks later to repeat the process. On one visit, each guy would get the AndroGel, which would crank up his testosterone to double the normal level. On his other visit, he'd unknowingly get the hand gel. This allowed us to directly compare the same guy as an Ordinary Joe and then as G.I. Joe.

When we looked at the results, we found that our testosterone-infused alphas were 27 percent less generous in the Ultimatum Game compared to themselves on placebo. But the underlying reason for this effect was not incidental; it's a chemical distinction with a purpose. It turns out that testosterone blocks the binding of oxytocin to its receptor, which puts the brake on the virtuous cycle I introduced in the previous chapter. The higher the testosterone, the more the oxytocin response is blocked, the less empathy the person experiences. The less empathy the person experiences, the less generosity.

So the empathy deficit we see in men isn't just a tagalong to being more aggressive. Testosterone *specifically* interferes with the uptake of oxytocin, producing a damping effect on being caring and feeling. At first this sounds like nothing but a negative. But by making young males—the hunters and warriors—not only faster and stronger but less nice, testosterone also makes them less squeamish about crushing skulls in order to feed and protect the family. Nice is often preferable, but when the job is to kill cute little animals for much-needed food, or to repel invaders trying to take your food (or your children), being overly "nice" is not so nice.

## Dealing with Loafers and Cheats

Not all threats in our environment of evolutionary adaptation came from the big fierce animals or big fierce neighbors outside the tribal compound. Because humans relied so heavily on a strategy of cooperation and group cohesion, survival itself was put at risk by any member of the group who didn't play by the rules—and not just those who were too aggressive. In a world in which filling your belly involved hard work and a fair amount of risk, the threat also came from anyone in the tribe who didn't do a fair share. Social scientists call this slacker behavior *free riding* or *social loafing*, and it's a serious problem. When the internal moral GPS within an individual fails, that's when the enforcement impulse in others needs to kick in. Our studies show that it's the testosterone that does the kicking.

By having testosterone block oxytocin's actions, nature saw to it that roughly half of the population would be moderately empathy-impaired, which meant ruthless, even unfeeling, when it came to punishment—no giving in to tears or excuses. Males became society's original enforcers for all matters large and small. Even today, when someone is blasting tunes on the beach, a lot of us will want to go over and explain to the perpetrator about the soothing sound of the waves (or at least about the virtues of ear buds). But very few of us will actually go over and have that conversation, and most of those who do are male.

Even chimps have an innate sense of how the game should be played. (If you doubt it, for the same task, reward one chimp with a grape and another with a cucumber and see what happens. Preferably from behind a glass partition.) Within the troop, a stingy chimp will get a seriously bad rep, so that next time there's some meat to pass around, the stigmatized-as-a-non-reciprocator will have to beg and solicit for a much longer time to have any hope of getting a share.

In our earliest Trust Game experiments at UCLA, long before we started infusing testosterone, we noticed that B-position players gave back a pretty consistent 41 percent of what they gained in the transfer from their A-player. But there was one significant exception. Whenever a male B-player received a transfer of anything less than 30 percent, he would hit back by returning next to nothing. The lack of trust implied by such a measly transfer really ticked these men off, because when we took their blood we found a spike in dihydrotestosterone, or DHT, the high-octane version of testosterone that stimulates ancient regions of the brain associated with aggression. DHT's effect on the brain is about five times larger than testosterone's. It not only unleashes aggression, but also increases dopamine, which makes the aggression feel good.

And there was a finely graded correlation—the lower the offers from A, the higher the DHT level for B—provided that the B-player was a male. This effect did not appear at all in women.

The women who received low transfers as B-players told me after the fact that they were "hurt," or "disappointed," and sometimes even "mad," but their anger never reached the point of prompting revenge. Instead, female B-players gave back a consistently proportional amount no matter how small the initial transfer they received.

This made me think about all the times I'd seen a woman speeding to catch up with a driver who'd cut her off, yelling insults and perhaps waving the middle finger. Which was never. And yet I see men doing this kind of thing all the time.

Men with high baseline testosterone have a hair trigger, and the desire to punish is automatic and emotional, blunt rather than nuanced, and blatantly reactive. This means that when testosterone is in charge, even ambiguous signals can lead to trouble. In traditional cultures—Sicily, let's say, or antebellum South Carolina—someone quick to take offense might be respected as a "man of honor." But the downside to this approach to life is, of course, keeping up with a heavy schedule of duels and vendettas that can get you killed. It also leads to a lot of teenage boys getting killed over the perception that somebody gave them a dirty look.

Evolution selected for this behavior—up to a point—because merely introducing the threat of punishment substantially increases pro-social behaviors, even when punishment is rarely or even never doled out. In psychological studies, the threat of punishment works even when the punishment itself is purely symbolic.

But nature gave the survival edge to groups with members who really *enjoyed* punishing the bad guys, even when punishment carried a significant personal cost, and brain scans have determined that punishing activates the dopamine-rich reward areas of the male brain much more so than it does in the female.

In our studies with the testosterone gel, our newly created alphas were *twice* as likely to punish others when they had their testosterone cranked than when they were operating on their normal hormonal level. To see how far these guys would go with this urge to hit back, we added an extra feature to the game: The B-player could not only withhold a reciprocal payment from the A-player who had seemingly dissed him, he could actively punish the guy by stating privately what he thought he deserved from the A-player. If the A-player's offer did not meet this unknown standard, all the cash on the line would be burned for both players. This allowed the B-player to, at a cost to himself, punish a stingy A-player with no benefit other than the pleasure it provided. A full 10 percent of our pharmacologically created alpha males chose to destroy all the available money rather than accept a low offer, while only 3 percent of unenhanced males exercised this option.

In fact, the guys on testosterone, when compared to themselves on placebo, consistently demanded more from others as a "reasonable offer." And both the stinginess and the punishment increased with a man's testosterone level. In another study, scientists inflicted mild electrical shocks. When men (but not women) observed non-cooperative partners getting zapped, they experienced not only activation of the reward areas of the brain, but a soothing deactivation of the pain matrix as well. So even if revenge does not literally taste sweet, it provides a warm balm of comfort— at least for men.

All and all, testosterone plus dopamine turns out to be the anti-HOME, a complete system of reinforcement for being "not nice." I call this testosterone-dopamine brain circuit TOP, for

Testosterone Ordained Punishment, which just so happens to be exercised quite often by males who view themselves at the "top" of the social ladder.

The benefit to the group in having at least some of its members wired to love punishing is that it reinforces morality by increasing the cost—as well as the likelihood of having to pay the cost—for anti-social behaviors. TOP is yet another counterargument to the idea that religious and other top-down moral invocations are the only way to achieve social harmony. Provided that we establish the right initial conditions, the system naturally creates its own incentives for playing by the rules, as well as disincentives for violating them.

Researchers at the University of Erfurt in Germany and at the London School of Economics did a study that powerfully demonstrated these natural effects. They used what's called the Public Goods Game, which involves setting up two investment clubs—A (the Freewheelers) and B (the Knucklerappers). The experiment tested eighty-four participants in thirty repetitions of either a two- or three-stage process, depending on the club. For both organizations, there was the decide-which-one-to-join phase, followed by the make-a-contribution-or-not phase. The difference was that among the Knucklerappers, but not the Freewheelers, there was a third phase for sanctioning or punishment.

To get the game going, each of the eighty-four participants received twenty euros. Each was then asked to join either the Freewheelers or Knucklerappers, and then to choose how much of the twenty-euro bankroll to invest. Whatever the participant chose not to contribute to the collective fund would go into his or her private account.

But here's the plot point that made things interesting: The only funds that would increase in value were those that went into the collective pot. And at the end of play, each club's pot would be divided equally among all of that club's members, regardless of their level of individual investment. In this setup, the most selfish free rider would receive just as much as someone who gave their all to expand prosperity for everyone.

At the end of the contribution phase of each round, all the players were updated on the contributions made by everyone else. They also were given the running score of their earnings to date, as well as everyone else's, in both groups.

Among the Freewheelers, with no third phase for sanctions, there was, in essence, a big WELCOME FREE RIDERS banner hanging over the door, and a lot of people accepted the invitation to try to get something for nothing.

Among the Knucklerappers, however, there was no place to hide. Each player was empowered to reward the generous, or nail the free riders, or both. Punishment took the form of a penalty token that could be assigned by anyone, and that would cost a designated non-contributing member three euros. But the pleasure of punishment was not free. The member who assigned the three-euro penalty would have to pay one euro himself. If you wanted to reward a generous contributor, the system worked the same way. A one-euro bonus token, assigned to a good guy, would cost you one euro.

At the beginning of the experiment, only about a third of the players signed up for the Knucklerappers. (Hey—I had my own experiences with nuns whacking us on the back of the hand with rulers. Who needs it?) And in the first round, free riders among the Freewheelers made out like bandits. They contributed nothing, but still shared in the increase from the collective pot.

After the fifth round, though, it was clear that the more demanding—and cooperative—gang at the Knucklerappers club was earning more. This realization had an amplifying effect, with more people catching on to a good thing. By the tenth round, seventy-five of the eighty-four players had changed their affiliation to Knucklerappers and embraced the idea of sanctions. With more members joining and contributing freely, the benefits of having an institution that clearly and evenhandedly enforced its rules became greater still. Think Switzerland, versus Mexico.

By the thirtieth round, the Knucklerappers were rolling in money, and everyone was contributing so generously that the need for sanctions actually faded away—the threat of punishment was enough. Meanwhile, the Freewheelers' assets had shrunk to zero.

Much of the credit for the Knucklerappers' high earnings belonged to those who bore the cost themselves to punish the free riders. In the first rounds there was no obvious benefit to dinking one another (other than the TOP circuit pleasure of retribution), but then the virtuous cycle began to roll, and it became obvious: An institution that promotes pro-social behavior not only by rewarding the good but by sanctioning the bad, delivers the highest returns.

If you doubt it, check out the value of Texas real estate, at least circa 2011, versus real estate in Arizona, Nevada, or Florida. Texans avoided the recent housing bubble fiasco by having and enforcing rules that kept people from turning the market into a casino. Today they're doing very nicely with solidly rising values. The other sunny places mentioned above embraced the casino concept and threw money at it like there was no tomorrow. Eventually, tomorrow came and they all went bust.

The upside to TOP is that it's an inducement for society's members to play by the rules. The downside, once again, is that it leads to young men killing one another because the killer felt "dissed."

It also leads to shouting matches over parking spaces, bar brawls, and lots and lots of domestic violence. High-testosterone males divorce more often, spend less time with their children, engage in competitions of all types, have more sexual partners (as well as learning disabilities), and lose their jobs more often.

So once again, balance is best. That's why nature paired testosterone (aggression and punishment) with oxytocin (empathy and cooperation) in a tag team, with the proportions of each allowed to fluctuate to suit the immediate circumstances.

## A Feminine Approach to Risk

The female side of the gender divide gave humanity closeness with the mother to prime the oxytocin receptors for empathy, the pair bond, and investment in children. Coming from the male side, there was enough oxytocin and empathy to participate in all of the above, but not so much that it interfered with testosterone-driven aggression, risk-taking, and rule enforcement. On top of this, there was the intermittent flush of testosterone (DHT) that fueled a genuine relish for punishing slackers and screwups, even when punishing came at significant personal cost.

But even though women have a stronger chemical bias for the pro-social behavior we now call moral, the evidence really does not support gender stereotypes. In some of our studies, the person with the highest testosterone was female, and gay men can be incredibly high in testosterone.

Naturally ramped up on testosterone, men have always been society's designated risk takers, encouraging the species to test the limits, and not just in the kind of stunts you might see on *Jackass*. A neuroscientist named Brian Knutson got guys to watch pornography to activate their TOP circuit, then asked them to choose one investment or another. The guys who'd been revved up sexually were 19 percent more likely to go with a high-risk investment than those who had not been admiring the ladies.

Back in the days of the wagon trains heading west, the men of the Donner Party chose to bet on an unproven shortcut. Unfortunately, that "quicker" route got them to the Sierras late, with snow coming on, which led to their being trapped in what's now known as Donner Pass. Half of them died; the others were reduced to cannibalism. Survivors' accounts indicate that the Donner women had been dead set against betting on the unproven route.

Risky business can indeed get you killed. On the other hand, too little tolerance for risk would have meant that all the pioneers—not just those on the Donner wagon train—would have stayed home and worked for wages back in Brattleboro (or Edinburgh for that matter) rather than becoming the boss of their own Ponderosa out west.

So once again nature gives us the yin and the yang, the antagonism between oxytocin and testosterone that helps to deliver the most sustainable balance, which is a trade-off between "Kumbaya" and a kick in the pants.

In our Trust Game studies, the only category in which women were more tight-fisted than men was when they were in the A-position and had to take a risk by transferring money in order to increase their return. The average amount they transferred on faith was about $4.50 out of $10.00, whereas men were willing to risk, on average, about $6.00. The greater risk aversion by women matches findings showing that women carry more life insurance, drive more safely, and invest their retirement funds more conservatively than men do.

There are solid, evolutionary reasons why women should be

more risk-averse than men. But is one approach or another—leading with oxytocin or leading with testosterone—better?

You never know till you get to the Sierras.

The highly risk-averse never start out, and the excessively risk-embracing often die along the way, which is why the give-and-take of a balanced gender perspective seems to deliver the best results over time. So maybe the Donner men should have listened more to their wives' misgivings. Or maybe they should have had a system like the one followed by certain Native Americans who have (a) male chieftains for each clan, but (b) male chieftains who can be removed by a vote of all the clans' mothers.

Trust itself is another area in which oxytocin can be tempered to good effect by a dose of testosterone. Having too must trust—remember the Pigeon Drop back at the ARCO station?—is another term for naïve, and naïveté can be just as deadly as being a dare devil.

At Utrecht University in Holland, researchers gave women small doses of testosterone, then asked them to judge the trustworthiness of faces depicted in photographs. On testosterone, the women were considerably less trusting than on placebo. Those most affected were those who had been the most trusting, which is to say socially naïve, before the testosterone infusion.

It so happens that the production of testosterone in women peaks just before ovulation. This increases libido just when conception is most likely to occur, but it also damps empathy just enough to increase wariness. Pregnancy and child-rearing require such a huge investment of metabolic resources, time, and energy that it pays for a woman to be skeptical—and choosey—about a potential mate's reliability. So nature balanced a drive to conceive and reproduce—friskiness at the most fertile time of the month—with a countervailing drive to avoid reckless choices.

## Winners

Humans are wired to be both trusting and skeptical, nurturing and punishing, competitive and cooperative, because each of these opposing forces can contribute to survival. But the biggest yin and yang may be the balance between competition and cooperation. It's testosterone that takes charge of the competitive part, and that's true for both genders.

Female college soccer players showed higher testosterone before taking on a particularly tough opponent, and when they won, their testosterone remained elevated for hours. By the same token, when you watch your favorite team lose a sporting event—even on TV—it reduces your testosterone regardless of your gender. If you identify heavily with a team (and here we have an obvious oxytocin influence), it makes you, by extension, a loser anytime they lose. So once again, we've got testosterone interacting with oxytocin—that empathy thing. But we're not just talking the Super Bowl or the World Cup. Even winning a spelling bee will cause your testosterone to go up, and losing will make it go down.

Everybody knows that a little competition can lead to a better performance. The moderate stress of moderate competition is actually good for us—it focuses attention, enhances memory and cognition, and provides clear goals. In moderation it also stimulates oxytocin release, which motivates us to draw on our social resources.

But women and men have different inflection points for which hormone will lead the way. The oxytocin effect is stronger in women. Under moderate stress, they more readily band together, a behavior that UCLA psychologist Shelley Taylor calls the tend-and-befriend response. In males high testosterone keeps the focus more on winning and letting the losers know they've been bested.

Winning too big too often can have a corrosive effect by bathing an individual in testosterone. Always coming out on top, consistently and over time, can reinforce some of the more obnoxious, stereotypically male behaviors associated with the hormone.

As we've seen, nice guys don't always finish last—amiability actually helps individuals rise in an organization. But then a curious thing happens. Being on top seems to turn people into jerks. Surveys of organizations find that most rude and inappropriate behaviors, such as shouting profanities, flirting inappropriately, and teasing in a hostile fashion, come disproportionately from those in the corner office. Achieving high social status appears to make it not just lonely at the top but morally perilous as well.

Which brings us back to alpha males and sex scandals. In 2011, when Dominique Strauss-Kahn, a prominent French politician and head of the International Monetary Fund, was accused of raping a hotel maid in a fancy New York hotel, it caused an uncharacteristic spate of soul-searching in France, where male privilege and pride in not being puritanical like the Anglo-Saxons is surpassed only by a cultural reverence for (and media deference to) elites. If you're witty and stylish and wield power, the normal rules that apply to *la bourgeoisie* simply don't apply to you. While the charges against Strauss-Kahn for the incident in New York were dropped, reports of other unwanted relationships with women have appeared. The French tolerance for extramarital affairs, combined with vast wealth and power, combined with a reluctance on the part of the French media—or anyone else—to exercise sanctions, created a brutally ugly scene. The event has also shone a light on the degree to which graft and corruption have been similarly swept under the rug by an atmosphere of mutual indulgence in which the powerful get a free ride.

Under the influence of testosterone, all leaders are likely to become more reckless and impulsive the higher they advance. In some cases, these high-testosterone habits can help an executive be more decisive and single-minded, or more likely to make choices that, despite being unpopular, will be innovative and profitable. That is, of course, unless and until the thrill of risk-taking, and/or single-mindedness in pursuit of an ill-considered objective, pushes the executive (and his company) into bankruptcy, or his army into a quagmire.

My studies using AndroGel showed that not only were testosterone-infused men more selfish, they, like Dominique Strauss-Kahn, also felt they were more entitled. Before the experiment began, I got each participant to tell me the smallest amount he would accept in the Ultimatum Game. These same men, after being pumped up on testosterone, rejected proposals that matched their own "acceptable" number 10 percent of the time. Men on placebo engaged in this inconsistency only 3 percent of the time.

People in power spend much less time making eye contact, at least when a person without power (which often means a woman) is talking. In tests like ours, administering testosterone has been shown to actually inhibit people's ability to pick up the social cues that eye contact conveys. This may be part of the reason that bosses ramped up on testosterone are more likely to rely on stereotypes and generalizations when assessing others and more likely to rationalize their own failings. After all, they're important people, with important things to do, and as they will readily tell you, they carry a heavy burden of responsibility.

In role-playing games, students pretending to be the boss were far less sensitive to the quality of arguments. It's as if it didn't even matter what was being said—their minds were already made up.

Deborah Gruenfeld, a psychologist at the Stanford Graduate School of Business, studied one thousand decisions handed down by the U.S. Supreme Court over a forty-year period and found that, as justices gained power on the court or became part of a majority coalition, their written opinions considered fewer perspectives and possible outcomes. (The scary thing is that the opinions reached this way, being *majority* opinions, are the ones that become the law of the land.)

Women's progress in business and politics has made being a jerk more of an equal-opportunity affair. It was a woman, the late real estate mogul Leona Helmsley, who famously said (shortly before being indicted), "Paying taxes is for the little people." Lifestyle mogul Martha Stewart couldn't bear to admit a minor infraction and accept a reprimand, so against advice of counsel she lied to federal investigators and wound up spending five months in prison. She then went on to compare herself to Nelson Mandela, one of the other "good people" forced to spend time in the clink. (Sorry, Martha—a little moral blindness there. Spending twenty-seven years in prison to win freedom for your people is a little different than going to prison for cheating on a stock transaction, then lying about it.)

Traditionally, men have prided themselves on being strong, silent types, and like Nelson Mandela, stoic in the face of hardship and pain. We also pride ourselves for "not getting all emotional" in positions of leadership. In the world of the traditional male— think Don Draper on *Mad Men*—cool and calm detachment is admired, and signs of emotion are stigmatized as weakness.

But women have long argued that *cool detachment* is sometimes just another way of saying "emotionally absent," meaning that by undervaluing empathy and intuitive perceptions, men miss the subtleties not only of eye contact but also of words, body language, and even social context.

In Stanley Kubrick's classic film *2001*, HAL the computer decides to kill all the humans on the spaceship he's guiding because he's calculated that their "feelings" will jeopardize the mission. In Vietnam, abstract reasoning (then secretary of defense Robert McNamara had gained fame for bringing quantitative management to the auto industry) led to using absurd "body count" metrics to justify tactics that made no sense. It also led to surreal statements such as "We had to destroy the village in order to save the village."

## Testosterone Teams

With diminished oxytocin and empathy, it's all too easy for the other to become The Other, and then The Enemy, and then The Inferior or The Demon. Without knowing the exact physiology, governments and armies realized thousands of years ago that the way to engage testosterone, decreasing empathy and increasing the desire to punish, is to manufacture an external threat to a group's existence. To give our hostility that extra oomph and wipe out any vestige of oxytocin's effects, it helps to then rebrand the opponents as monsters.

The ancient Greeks and Persians trash-talked each other as barbarians. In modern times propagandists came up with terms like *Yellow Peril, Axis of Evil, Evil Empire, Hun, Kraut, Jap, Red,* and *Commie* to cut off any notion that the person being opposed had human qualities and, sometimes, valid enough reasons for doing what he was doing other than that he was subhuman or possessed by Satan.

Economics Nobel laureate Vernon Smith showed that simply using the word *opponent* rather than *partner* was enough to cut levels of trust in half. When the word *partner* was used to describe another person, trust registered at 68 percent. In the same situation, when the word *opponent* was used, trust dropped to 33 percent.

Sometimes a reliance on rational abstraction to the exclusion of empathy contributes to a completely irrational deference to authority. In the early 1960s, psychologist Stanley Milgram did a famous experiment in which he asked people to administer a mild electric shock to another person hidden from view, but seated close enough to be heard. A lab-coated scientist—the "authority figure"—kept asking the person administering the shock to give it just a little more juice. The participants were incredibly compliant, even after they began to hear screams of pain. (Neither the pain, nor the screaming, nor the electric current was genuine, but the participants were told that it was.) Two-thirds of participants zapped their neighbor with what they thought was extra voltage even when they were advised that it might be fatal. After all, an authority figure was telling them to do this, so morally—at least as they saw it—they were off the hook.

In another classic study called the Stanford Prison Experiment, psychologist Philip Zimbardo randomly assigned volunteers to the role of guard or inmate inside a mock prison that operated around the clock. After six days the prison atmosphere had become all too real, and ugly, and the whole thing had to be called off. The guards became sadistic and tortured the inmates. Some inmates became passive and accepted the abuse. Others all too readily complied when asked to inflict punishment on other prisoners. Even Zimbardo himself went over the edge in his capacity as "prison superintendent," losing sight of his role as a psychologist and letting the abuse get out of hand.

In-group/out-group distinctions can trump empathy and lead to very bad things in part because when we follow the crowd, the dopamine system kicks in, which makes groupthink and conformity pleasurable. (The flip side is the pain we feel anytime we're excluded from a group or a relationship. It turns out that the brain processes social pain exactly as if it were physical pain.) This pleasure/pain, push/pull reinforces group identity, even when the group turns into a lynch mob.

Testosterone signals the in-group/out-group distinction even in a rivalry as benign as a game of dominoes. In a study done in the Caribbean, testosterone levels were much higher when men played opponents from a neighboring village than when they played a competitor from their own village. Psychologists have shown that any random assignment to sides—"You guys over here are the redbirds, you over there are the bluebirds"—is enough to spur in-group/out-group competition.

Put all these forces together—high testosterone, deference to authority, group pressure, dehumanizing abstractions—and you get the insanity of the Nazis during the 1930s and '40s, or the Belgians in the Congo in the late nineteenth century, who punished workers for low yields on rubber plantations by cutting off the hands and feet of their children. More recently, we've seen mass executions, rapes, and mutilations during genocides in the Balkans, Rwanda, and Sudan, and even in the war among rival drug cartels taking place along the U.S.-Mexico border. If you're not one of us, you deserve to die—and the more cruelly the better.

We have the most empathy for those close to us, but when we're threatened our brains make a very simple us-versus-them calculation. Is this person part of my group or from another group? By damping oxytocin, the stress of fear narrows the circle of empathy, while also limiting our focus to a very amoral calculation of what we need to do to survive.

In *Jarhead*, Anthony Swofford's memoir of the 2003 Gulf War, he describes being "afraid of the enemy's humanity" because seeing them as human makes it difficult to pull the trigger. Killing even a dehumanized enemy can itself be dehumanizing—which can be a contributing factor to what's come to be known as post-traumatic stress disorder. In an effort to rehumanize themselves, soldiers exchange hugs, back rubs, and say a lot of "I love you, man." Each of these gestures releases oxytocin and reanimates the Moral Molecule, reducing stress and bringing the men, who've been forced to do some very inhumane things, back into the human family.

The poet and spiritual leader Robert Bly spent years on the lecture circuit making the case that you can't improve male behavior by shaming men or by trying to turn them into women. You have to honor the virtues of testosterone—what the ancient Greeks called *andreia*—while making sure that the human being with all that aggressive juice also has a fully integrated head and heart.

In that context, especially, gender equality is a great thing. It can make life less stressful for everyone: women not being constrained by traditional roles, and men not feeling that they have to hold all their feelings in or walk around with a chip on their shoulder. Ideally, both genders can share the burdens and the joys of living all of life, and not just one half or the other, while also appreciating some essential differences.

But there's no doubt about it—testosterone is the problem child when it comes to pro-social behavior. It declines naturally in men beginning around age thirty, which makes older men less aggressive and more empathic, so the age-crime pattern reverses as men enter the andropause phase, the male analog of menopause. Around age thirty, the prefrontal cortex in men is at long last fully wired up, which allows the executive brain to do a better job of inhibiting impulsivity, which leads to greater deliberation.

Testosterone effects are also diminished when a man commits to a woman, and if that relationship produces children, testosterone declines even further. In my own case, I joke that I'm a "girly man" because I have two daughters and spend a lot of time brushing their hair and picking out dresses for them. Back when I was a teenager working on cars and playing football, I never imagined I'd be doing this sort of stuff, but now I love it. I now take fewer risks (occasional skydiving episodes notwithstanding), get into fewer confrontations with other men, and drive more carefully. I also think I've become a lot more tolerant and forgiving.

Another benefit to having males with greater empathy doing more to nurture their children is that it adds to the loving attention kids need in order to develop the oxytocin receptors necessary to become fully empathic human beings.

Unfortunately, the reverse is also true. Dysfunctional parents tend to be empathy-impaired, which often leads to stressed and traumatized children who, in turn, grow up with insufficient oxytocin receptors, which perpetuates a vicious cycle of empathy-impaired people.

# CHAPTER 5

# The Disconnected

*Victims of Abuse, Bad Genes, and Bad Ideas*

A few years ago, at Loma Linda University Medical Center, my wife ran an outpatient epilepsy clinic that had a large number of trauma victims. Many were women who had been abused so badly that the emotional damage began to manifest itself in physical symptoms—a process called somatization. They became convinced that they were paralyzed or having seizures or heart attacks.

In animals, extreme neglect or abuse can shut down the physiology of connection that oxytocin release makes possible. I came to the clinic to see if we could use the Trust Game to find evidence of a similar effect in humans.

The first patient we studied was a twenty-two-year-old named Alicia who had been raped repeatedly by her stepfather, beginning when she was twelve. Even though there was nothing physically wrong with her legs, she arrived in a wheelchair. When the doctors encouraged her to stand she could, and she was able to move along at a shuffling gait, but in her mind she was completely paralyzed.

Alicia was in heartbreakingly bad shape. As we took her blood I noticed the way her head drooped down, and that she didn't make eye contact. She cooperated fully, though, and she played the Trust Game just fine. But when she was a B-player and received a transfer of money, the display of trust did not cause her oxytocin levels to spike. Just as we'd anticipated, her childhood trauma had shut down her oxytocin receptors.

It took us a year to run fifteen of these somatization patients through our Trust Game protocol. They all had very unstable lives and simply wouldn't show up when they said they would, or they'd move or change phone numbers, which made it very hard to keep track of them.

We later did a brain imaging study on these patients, showing photographs of people in distress that usually arouse empathy. For these trauma victims, the amygdala was disengaged. It has a high density of oxytocin receptors and modulates emotions. So just like Alicia in the Trust Game, they were emotionally flat, and they showed no response to the photographs.

When we tried to reach Alicia again to ask her to come in for a brain scan, the relative who answered the phone told us she was dead. Then she hung up before we could offer condolences, or even find out what had happened.

## The Five Percent

We had found evidence of oxytocin impairment in our very first Trust Games at UCLA. The last player during that initial series of tests was on the heavy side, so much so that my physician buddy Bill Matzner had to stick him four times to get through to a vein. After we got the young man's blood, I apologized for all the pain we'd inflicted.

"No problem," he said. "I love this experiment. Can I come back tomorrow?"

This made me curious—most people who get stabbed three or four times for a single blood draw are not very happy about it. So I did a little checking.

Turned out this fellow had been a B-player. The A-player he'd been paired with had transferred every cent of her $10 endowment over to him, which, multiplied by three, gave him a go-ahead of $30, which when added to his initial endowment of $10 gave him a total of $40. But even though the A-player had bet it all to increase the size of the pie for both of them, this guy kept every cent.

In behavioral economics, the technical term for people like this is *unconditional nonreciprocators*. In my lab we call them *bastards*.

Over time we discovered that 5 percent of our mainstream, college student volunteers were like this. They chose not to give anything back no matter how much money the other player trusted them with. In these cases, trauma was not a factor—they were all high-functioning college students who had not experienced severe trauma. When we analyzed their blood, we found that members of this 5 percent actually had an *excess* of oxytocin. This seemed counterintuitive at first, but not when we considered the fact that the system does not react to the overall level of oxytocin but only to an immediate surge. The off-switch in their receptors was malfunctioning, flooding the system with oxytocin, and the overflow was creating a functional deficit. No surge, no contrast, no oxytocin activation. No oxytocin activation, no empathy or reciprocity. So even though the problem for these unconditional nonreciprocators begins with too much oxytocin, we call the condition Oxytocin Deficit Disorder because they simply don't release oxytocin when they should.

Over time, we determined that there are actually three broad categories of influence that damp or destroy the oxytocin response: temporary, acquired, and organic.

Anyone can have a bad day, and transitory job worries or a bad commute can damp the oxytocin response. Victims of trauma like Alicia represent acquired oxytocin impairment at one extreme, while people at the opposite end of the social spectrum can lose their ability to empathize via elevated status or mental rigidity. Organic impairments include a number of genetic conditions, with autism being the most familiar, and with psychopathology being the most extreme.

## A Little Stress Can Go a Long Way

Having a bad day comes in all shapes and sizes, and the most common culprit in creating an oxytocin deficiency is stress, which from time to time can make any of us a temporary non-reciprocator. And we don't have to be under rocket fire, or desperate to find a job, or worried about a child in the hospital for stress to damp oxytocin release and make our behavior less generous.

There are two basic types of stress—chronic and acute—and both interfere with the HOME system. Stress of the "rocket attack" variety causes you to release the hormone epinephrine, also called adrenaline, which prepares you for fight-or-flight. It's a familiar tale by now, but epinephrine raises your heart rate and blood pressure and makes you breathe more rapidly. High levels of this stuff can even make you vomit, or void your bowels and bladder. This response came in handy when our ancient ancestors needed to lighten their load while running away from predators, but it's not particularly helpful when the cause of the upset is turbulence on an airplane or a crucial disagreement with your boss.

Most people in the modern world spend a lot more time dealing with stress that isn't big and dramatic but simply woven into their daily lives. This variety, called chronic stress, causes the body to release a chemical called cortisol. Given that our bodies evolved to fit our environment of evolutionary adaptation, which was the plains of East Africa several million years ago, this hormone also served to help us escape from threats, but in a more sustained way. Epinephrine is the immediate jolt to action, but then cortisol comes in to keep the elevated heart rate elevated, the higher blood pressure high, and the accelerated breathing accelerated, just as if your village was caught in a flash flood and you had to work for hours to get your children to safety. It also liberates glucose from fat stores so your muscles will have energy ready to burn. The downside of this once very useful adaptation is that when it's triggered by today's persistent low-grade anxieties, these high-heart-rate, high-blood-pressure, high-glucose responses stay with us and become toxic in all sorts of ways, causing heart disease and diabetes as well as the impairment of moral behavior.

High levels of epinephrine and cortisol both inhibit the release of oxytocin, which throws a monkey wrench into the virtuous cycle to diminish empathy and shove your active concern for others right out the window. The evolutionary logic here is the same as the rationale behind those airline safety instructions that tell you to put on your own oxygen mask before trying to help your child. When you're struggling to survive the next sixty seconds, a high degree of altruism, or even a refined sense of moral scruple, may not be the best way to go.

And it doesn't necessarily take all that much stress to make us less virtuous. A study with seminarians found that a high percentage of even these especially altruistic and spiritually committed young men failed to stop and render aid to a moaning homeless man when they were late to class.

The more pernicious effect of too much cortisol from too much stress is that it can lead to longer-term empathy fatigue. I think this happens to a lot of people when the hassles of modern living combine with overexposure to the news media. There's so much strife and hardship in the world, and so many people and causes crying out for attention, that sometimes we just want to curl up in a ball. The constant cascade of stressful stimuli can cause us to burn out, just as too much front-line duty causes emergency care physicians and first responders to burn out.

Some people are more resilient; for others, a little stress goes a long way toward empathic exhaustion. For some, just the stress of feeling isolated can *feel* like a life-threatening experience, which, for our primitive ancestors, it once was. So even loneliness can damp down HOME and make us less nice, and less caring, just when we really need to reach out. But the emotion that's the most toxic of all is hostility.

In the 1950s a San Francisco cardiologist named Meyer Friedman had set out to redecorate his waiting room when his upholsterer noticed that the chairs showed wear only on the edges. He mentioned this to Dr. Friedman, and in a flash the good doctor hypothesized that his patients were so anxious that they were literally sitting on the edges of their seats. So maybe their anxiety was contributing to their heart problems? He began studying this kind of behavior and came up with the concept of the "Type A personality," which had a lot of influence in the context of risk factors for health. It also created a lot of controversy and has been badly misconstrued.

Type A people are the hard-driving, impatient go-getters, but the fact is that millions of Type As actually live happily ever after—and in good health—working long hours, racing to meetings, and yelling into their cell phones. The critical issue in matters of stress, health, and oxytocin release is the happiness part.

If you love being a Hollywood agent, an investment banker, a politician, or a star sales rep, and if you're on top of your game and you're getting all the breaks, your Type A life is crazy busy—but not necessarily all that stressful in a corrosive way. It turns out that what unleashes the deadly effects of cortisol is bottled-up anger, the kind that comes from being frustrated and socially subordinated. The people in the greatest danger from heart attack or stroke, as well as from Oxytocin Deficit Disorder, are not necessarily those who're racing to the top, but those who've gotten stuck in the middle.

In England epidemiologists carried out a lengthy examination of the British civil service called the Whitehall Study, and its findings demonstrated that the worst position to occupy in terms of health and well-being is a job with high responsibility and a low degree of control. In these mid-level positions, the switch for the supply of high-stress cortisol often gets locked in the on-position. This is the same squeeze in which we find doctors who feel hamstrung by insurance regulations, or teachers caught between parents who think their little Johnny can do no wrong and school administrators who won't enforce standards and discipline, and a lot of people stuck selling tickets or sweeping up after, when what they desperately want is to be up in the spotlight, chewing the scenery.

What makes this kind of social stress especially problematic for society is that it's now endemic. Job tenure is getting shorter and the anxiety of finding the next place to earn a paycheck can cause chronic stress. So while society's winners are having their oxytocin blocked by a flood of victory-induced testosterone, those who aren't doing as well can become empathy-impaired as a result of their anger and frustration at being hemmed in.

American mythology has blue-collar guys down as being more manly, but in a society where money and status are equated with human worth, being in a subordinate position can be humiliating. This may be one of the reasons that you see so many lower-income guys pushing the hostility outward by trying to look menacing, whether it's in black leather vests on Harleys or with shaved heads and tattoos. The same is true for young guys from the inner city with their hoodies and their shades and their pants drooping below their butts, prison-style.

When the humiliation of low social status combines with real economic insecurity, the sense of being squeezed can boil over into a rush of oxytocin-impairing DHT. This is probably one of the reasons our political discourse is so polarized these days. Anger and lack of empathy create a negative loop in which it's all too easy to simply lash out and blame "the other," whether "the other" is illegal immigrants, "those fundamentalist morons," or "elites." Meanwhile, the "elites" seem utterly tone-deaf when it comes to anticipating how their actions—the Wall Street bailout; corporate execs coming to Washington in private jets to plead for federal help—are going to be viewed by the average citizen.

## Scarred for Life

The more entrenched problem for millions of people like Alicia, in our opening story, is that their empathy impairment is not a transient condition but deeply embedded as the result of serious emotional scarring. The primary culprit here is abuse—which can include neglect and abandonment—in early childhood. When kittens are deprived of light, the visual areas of the brain atrophy. In the same way, if the oxytocin receptors are not stimulated by love and attention early on, they fail to develop.

In 1958 psychologist Harry Harlow conducted an infamous experiment in which he took newborn rhesus monkeys away from their mothers. He then presented them with two surrogates, one made of wire and one made of soft cloth. Either stand-in could be rigged with a milk bottle, but regardless of which "mother" provided food, infant monkeys spent most of their time clinging to the monkey mannequin made of cloth, running to it immediately when startled or upset.

It turned out, though, that just as mothering wasn't only about food, nurturing wasn't just about making a baby feel good in the moment. Later on, the monkeys with wire mothers showed significant delays in their mental and emotional development. Monkeys raised in isolation fared even worse. Even after they were allowed to rejoin their monkey troop, they'd sit alone and rock back and forth. They were overly aggressive with their playmates, and in adulthood they failed to form normal attachments or develop even the most basic social skills. When an ovulating female who had been socially deprived, which is to say oxytocin-deprived, was approached by a normal male, she'd squat on the floor rather than present her hindquarters. If a previously isolated male approached a receptive female, he might clasp her head instead of her hindquarters and start thrusting.

Females raised in oxytocin-deprived environments became either incompetent or abusive mothers themselves. Even monkeys raised in cages where they could see, smell, and hear—but not touch—other monkeys developed social withdrawal and rocking as well as excessive grooming and self-clasping.

Harlow's experiment was brutal, but it might have helped avert what amounted to a humanitarian disaster if only its lesson had been more widely absorbed. Instead, in the decades that followed, thousands of orphan children were consigned to an emotional gulag, most notably in Romania, where the system established by former dictator Nicolae Ceaușescu provided one caretaker for every twenty infants. This meant that there was barely time for basic hygiene. Hugs and other loving attentions were out of the question.

When the orphanages were opened to the world in 1989, outside health officials found three-year-olds who did not cry and did not speak. These children were grossly delayed in physical growth, motor skills, and mental development. Like Harlow's oxytocin-deprived monkeys, they clutched themselves and silently rocked back and forth.

American scientists later studied some of these orphans who'd been adopted. They had them play with their adoptive parents, then collected their urine. Even after three years in a loving home, the children who'd been emotionally deprived as infants showed no rise in oxytocin after thirty minutes of parent-child play.

As we saw at the clinic with Alicia and the others, impairment from earlier deprivation can be permanent. At Loma Linda we used the Trust Game to compare victims of chronic childhood sexual abuse with women who'd experienced healthy childhoods. Average trustworthiness was not hugely different among abused women next to those who'd been more fortunate—abused women returned 49 percent while controls returned 53 percent. But when we looked beyond those raw statistics, the abuse victims showed a much wider range of behaviors, and therein lies the tale.

Twenty percent of the abused women were extremely trustworthy (returning more than two-thirds of the money they acquired), compared with only 2 percent of the control group. On the other end of the spectrum, 33 percent of abused women were very untrustworthy (returning less than one-third of the money they controlled) compared with only 12 percent of the women in the control group.

What these disparities tell us is that the HOME system in these women was fundamentally dysregulated, which left them emotionally disengaged. Our brain function study of them showed the same thing.

Being trusted with money was associated with a 1 percent increase in oxytocin for the victims of abuse whereas the control group had a 7.5 percent average increase. Oddly enough, the higher the oxytocin release among abused women, the less trustworthy they were.

The abused women reported nearly half as many close friendships as did controls, and they were far more likely to avoid romantic relationships (60 percent avoidant compared with 20 percent in controls). The abused women also had 36 percent lower baseline levels of cortisol compared with the controls, which reveals another source of their emotional flatness.

## The Goddess of Greed

In 2011 a film crew came to our lab to do a documentary on the Seven Deadly Sins, and they brought along an unusual young woman named Stephanie Castagnier. If you watched Donald Trump's *The Apprentice,* you may remember her as the Canadian-born real estate wizard who presented herself as, essentially, the goddess of greed. She's very attractive, very feminine, but with the aggressive drive to make her first million in a bare-knuckles business before the age of thirty.

I designed a set of experiments to examine her HOME system. Despite all her aggressiveness in business, it turns out that she has unusually low testosterone. But she has a genetic anomaly in which her body manages to extract incredible amounts of DHT—the high-octane stuff that often prompts punishing behavior—out of the limited testosterone she has to work with. DHT, of course, blocks oxytocin. So Stephanie is, as she freely admits, like a lot of guys in being incredibly driven and not so good at empathy. But there's more to her story.

When she was a child her father had been a high-rolling drug dealer, so while her family had plenty of money, she lived in a kind of war zone with machine guns under the bed, cash stashed in pillowcases, and strange—sometimes violent—goings-on at all hours. By the time Stephanie was in middle school, her father's own drug use had knocked him down to street dealer, and ultimately to homeless junkie. During this period he would steal her sneakers, her jacket, her textbooks—anything he might be able to sell for a few dollars to spend on drugs. She slept with a baseball bat beside her bed for fear that he might sell her to one of his junkie friends, or that someone would simply break in and rape her. Both her parents died from AIDS before she finished high school.

We had Stephanie watch the video about Ben, the little boy with cancer, and she told us she felt moved and that she was working hard to hold back the tears. But her blood tests showed no oxytocin release, which means no real empathy. Being a highly resilient and savvy survivor, she simply knew how to say all the right things.

When I called Stephanie to share her test results with her, I cautioned that she might not want to know what I'd found; it might be too revealing of her inner self. She said she did want to know. One implication of ODD is an inability to sustain romantic relationships. Stephanie laughed and said she went through men like so many pairs of running shoes.

## Autism and Anxiety

For millions of people, ODD is not the result of early life experiences but the genetic cards they were dealt. The most prevalent disorder in which oxytocin impairment is thought to come into play is autism. One study found that children with autism have lower baseline levels of oxytocin in their blood. Autistics also have lower levels of oxytocin in their spinal fluid, which suggests that the oxytocin-producing neurons in the hypothalamus don't function properly. Other studies have found variants in the oxytocin receptor that may prevent it from properly binding with the hormone. So far there is no conclusive evidence that amounts to a "cause" for autism, but research on prairie voles shows that when their oxytocin receptors are blocked it prevents the animals from forming normal social attachments. So the inference is pretty strong.

A high level of fetal testosterone has also been implicated in shutting down the HOME system in autistics. Some experts go so far as to call autism "extreme male brain syndrome," and it's true that among the six people per thousand in the United States born with the condition every year, there are four times as many boys as girls.

What we know for certain is that autism affects communication skills, the ability to read others' emotions (in other words, empathy), and the ability (and/or desire) to connect socially. Repetitive and stereotypical patterns of behavior are also associated with the condition, and these include the kind of rocking we saw in those monkeys deprived of close contact early in life. It's usually these behaviors that lead to a diagnosis of autism, and most often the diagnosis is made by about the age of the three.

Even so, the range of impairment is so broad that the term most often used is *autism spectrum disorder*, with Asperger's syndrome being the mildest form. Asperger's allows people to get along well enough within social groupings, often while functioning at an exceptionally high level in areas that require technical expertise. Some experts argue that the list of notables with Asperger's includes Isaac Newton, Thomas Jefferson, and Albert Einstein. It's been suggested that Bill Gates, known to rock back and forth during tense meetings, might be included. There's even speculation that the computer industry would never have achieved what is has without the contributions of thousands of high-functioning Asperger's people who would rather write code than socialize. (There's further speculation—and it's only speculation—that double-Asperger's marriages within the industry have caused an epidemic of autism in Silicon Valley and other high-tech centers.)

Not surprisingly, autistics don't behave the way other people do when playing social scientists' favorite games. In one study involving the Ultimatum Game, 28 percent of autistics offered nothing. Among controls, only 3 percent made this choice.

The fact that offers of nothing are always refused—they're a waste of time, in other words—suggests the gravity of the impairment. It's very hard to get along in our society without a high degree of social savvy. By the same token, autistics have a tendency to accept low offers because their Theory of Mind misses the subtleties of give-and-take, which is the essence of productive cooperation.

Lisa Daxer, a biomechanical engineering student at Wright State University, writes a blog called *Reports from a Resident Alien* in which she refers to non-autistics as "neurotypicals." She also expresses her amazement at how they (which is to say, all the rest of us) are fixated on social interaction. She writes about watching her friends watch *Friends* (a very "neurotypical" show, she calls it) and describes them mimicking the expressions on the faces of

Jennifer Aniston or Courtney Cox. "You have to actually *interfere* to stop neurotypicals from socializing," she told NPR, with some amazement.

Lisa Daxer has few if any of these social impulses, but being highly intelligent and a gifted problem-solver, she also realized her need to work on developing social skills. After all, no matter how brilliant your engineering, you don't get a spaceship to the moon all by yourself. Science and technology—like almost everything else—is done in teams, which requires empathy, Theory of Mind, and the ability to use co-cognition in addressing common goals.

Like another accomplished autistic, the animal scientist Temple Grandin, Lisa achieves through intense cognitive effort what the rest of us manage to do intuitively. Don't stare. Take turns. Don't stand too close. (In the interview with neurologist Oliver Sacks that first brought attention to Temple Grandin, she described how being among non-autistics made her feel like she was "an anthropologist on Mars.") Daxer also set out to memorize a list of topics that were off-limits, which she says includes sex and "anything that happens in the bathroom." She finds some of these neurotypical taboos odd, but then she compares them to her aversion to certain fabrics. "I avoid polyester clothing," she said. "They avoid talking about death."

I've had dinner with Temple Grandin several times, and the first time I met her she appeared so fragile that I instinctively put my hand on her arm. Then I remembered that most autistics don't like to be touched. I found out later that at eighteen, she actually created a squeezing machine she could climb inside to approximate a stress-relieving hug without having to interact with people.

At dinner Temple could hold a conversation and make eye contact, but her facial expression was wooden, without any hint of emotion. After dinner she skipped dessert and was the first to leave.

Could oxytocin be used to help people like Lisa Daxer and Temple Grandin relate to others more easily?

There's a survey called the Autism Spectrum Quotient test, which uses fifty questions to measure social behaviors, the ability to read others' emotions, and the need for routine. In one study conducted at Mount Sinai School of Medicine in New York, twenty-seven men were rated on this scale, then infused with oxytocin. Afterward they were asked to watch a video of people interacting and discussing emotional events. They were also asked to rate the emotions on display. For those who received the oxytocin infusion, emotional accuracy increased (compared to themselves on placebo), but this was true only for those with the highest autism scores. In other words, oxytocin did not turn people who were already socially adept into even more masterfully empathic and socially savvy creatures. It simply helped those who needed help the most.

This study suggests that even those with the most serious deficits have some intact oxytocin receptors that can be engaged through social interactions. Flooding the brain with oxytocin by way of an inhaler has been shown to help autistics to decrease self-soothing behaviors such as rocking, to moderately increase eye contact, and to pick up emotional cues in speech. But a major component of the HOME system is the release of serotonin, triggered by the release of oxytocin. Serotonin is, of course, the neurotransmitter that lowers stress and gives us an overall sense of well-being. So the benefit to autistics from oxytocin infusion may be that it simply lowers their typically high levels of anxiety.

For several reasons, I don't think oxytocin infusion will ever be a realistic therapy for autism. For starters, the experience of infusion sucks. Second, the effects last only for a short while (though a longer-acting oxytocin formulation called carbetocin is currently in clinical trials). Mostly, when the problem is a shortage or malfunction of oxytocin receptors, increasing the concentration of the hormone alone isn't going to do much good.

A more promising tack is to increase the number of oxytocin receptors. Nearly all the neurologic and psychiatric patients I've tested release oxytocin, even if it's just a little. Having more receptors means having more places for the molecule to bind, which means making the best use of the oxytocin that's already available. This approach has been proven effective in rodents, and it's now moving into clinical trials with humans. If it's proven effective in humans and gains FDA approval, it might be able to help with disorders ranging from autism to social anxiety to post-traumatic stress.

The distinctive advantage of relying on increasing the number or sensitivity of oxytocin receptors is that it maintains the integrity of the Moral Molecule's function in regulating social behavior. In other words, it can switch on and off. Increasing the number of receptors to increase social engagement allows the oxytocin effect to come into play the normal way, which is when oxytocin release is prompted by appropriate social stimuli, such as signs of trust or affection. Simply flooding the brain with oxytocin via an inhaler is more like flooring the gas pedal on a car—not very subtle, or sensitive to outward circumstances. With a simple pedal-to-the-metal oxytocin infusion, patients might become so trusting that they'd be easy marks for the Pigeon Drop as well as more damaging forms of victimization.

Imbalances in oxytocin may also play a role when the problem isn't the inability to connect with other people, but simply tremendous anxiety about doing so. Recently I helped run an experiment at Massachusetts General Hospital in which patients with social anxiety disorder (SAD) played the Trust Game. When these people were in the B-position, they gave back 6 percent less than the people in the control group, who had no symptoms of SAD. This finding jibed with the fact that the SAD patients also had a much higher baseline level of oxytocin, meaning that, once again, their system was already flooded with oxytocin, which prevented a surge in response to stimulus. For the SAD patients, the higher the baseline oxytocin, the higher their reported level of dissatisfaction with social relationships.

There's been talk of a medication that targets the HOME system in an effort to relieve this kind of anxiety, but my preference remains hugs not drugs.

A few years ago a woman from the United Kingdom read about our research, then tracked me down to ask about her daughter. The young woman was having panic attacks whenever she was in a group setting, which included the offices of the large corporation where she was an executive. I described how petting a dog, getting a massage, and being around people who projected a high degree of trust might be able to help. If that didn't do the trick, I recommended seeing a psychiatrist who might prescribe one of the anti-depressants—such as Prozac or Paxil—that at least in rodents increase oxytocin release.

The young woman's mother told me that the only time her daughter felt comfortable around another person was when she was giving a massage. Which is, of course, another way of self-medicating, because even *giving* a massage can cause the brain to release oxytocin, and this can train the brain to increase oxytocin release that can lead to greater ease in social interaction. Massage is also a direct attack on the stress of social anxiety, because, once again, serotonin triggered by oxytocin release is a relaxant. Eventually, the young woman quit her corporate job and became a massage therapist.

For children, a certain amount of social anxiety around strangers protects them, but eventually, with adult guidance, the child needs to learn to tune the system. Uncle George or neighbor Sue or teacher Ann should be treated like family, parents explain, but the strange man at the mall who wants you to come with him to the parking lot—no way.

Parental lessons, along with the child's own life experience, bring the cognitive part of the brain into play to help the child learn to regulate trust and distrust. When children begin their lives with loving attachments to adult caregivers, the oxytocin system rapidly develops, facilitating the kind of reciprocal affection and caring that serves us well in adulthood. You be nice to me and I'll be nice to you.

## Too Much Trust

The opposite of social anxiety is having no social boundaries whatever, a less common problem, but one that offers yet another window on the workings of oxytocin. In the lab of neuroscientist Antonio Damasio, I came across a study participant who had Urbach-Wiethe disease. This genetic disorder damages the amygdala—the center of wariness—but leaves the rest of the brain intact. Ms. Smith, as we'll call her, was completely open to other people—hyper-gregarious, in fact—but she had no ability to size up the moral character of anyone she met. She could detect happiness and other emotions in faces, but not threat or danger. She was also unable to pick up the subtle facial cues that allow us to determine which individuals are not trustworthy and best kept at a distance.

When you first meet Ms. Smith, she comes off as a bit disheveled but, aside from a failure to maintain eye contact, pretty normal.

But then she gets too close, both literally and figuratively, failing to maintain the usual social space between herself and others that most people intuitively recognize. And then the personal information starts to flow, with details that most of us would be embarrassed to tell a doctor or a therapist. She doesn't seem to notice people recoil as she describes her sex life or the details of a recent medical exam. More to the point, she does not vary her behavior as she interacts with different people: friends and strangers, old and young, kindly or malicious—she treats everyone like a best friend from way back.

Ms. Smith has the cognitive ability to live on her own and take care of her own affairs, but her impairment makes her a super-truster, which often leads to victimization. This lack of skepticism and judgment of character may also help account for why she has three children by three different fathers.

We weren't able to measure Ms. Smith's oxytocin levels, but we had her play the Trust Game four times. The first time around she was very trusting as a player A, but then as a player B she did not reciprocate. The second time she was a player A she reversed her behavior and transferred almost nothing to the B-player she'd been paired with. So it seemed that she was in the same boat as the trauma victims like Alicia—her trust/reciprocation governor simply didn't work properly.

Urbach-Wiethe disease is exceptionally rare. A more common genetic disorder that causes people to become super-trusters is Williams syndrome, which affects roughly one out of every ten thousand babies born in the United States. (Still pretty rare.)

Williams patients have no social fear, so even as toddlers they run up to strangers and make intense eye contact. Like Ms. Smith they will invade your physical and emotional space, but in a very loving way. All through life they are hyper-responsive to the smallest opportunity for social interaction, chatting away with total strangers.

When most people are exposed to pictures of faces expressing fear, their amygdala activates, but Williams patients show no response. The amygdala is an area of the brain with a high concentration of oxytocin receptors, so it's possible that oxytocin plays a role in this syndrome, but at this point we just don't know.

One clue to the origins of the disorder is that Williams people are missing a suite of genes on chromosome 7. Some of these genes are expressed in the hypothalamus and in the pituitary, brain regions that produce and release oxytocin. These genes also alter eye movement, which may account for the intense eye contact and focus on other people common to Williams patients.

Could some form of oxytocin therapy help in any of these conditions?

In animal experiments, oxytocin infusion has been shown to ease the symptoms of withdrawal from heroin, cocaine, and alcohol, but as of now we don't know if this will work in humans. If it does, the fact that oxytocin release also triggers serotonin release could provide a second dose of healing in the form of anxiety relief.

Another promising chemical intervention is the use of the mood stabilizer lithium, which also appears to increase oxytocin. In 2009 Japanese researchers reported that suicide rates were lower in areas in which naturally occurring lithium was in the drinking water. (From 1929 through 1950, lithium was an active ingredient in a lemon-lime-flavored soft drink called Bib-Label Lithiated Lemon-Lime Soda. In 1936 this drink changed its name to 7UP.)

Recently MDMA, the drug commonly known as ecstasy, was shown to cause oxytocin release, which probably explains the drug's I-love-everybody effect on users. Recent studies with hospice patients have shown that ecstasy can lessen anxiety and smooth the way to better social interactions. Unfortunately, even a very few doses of MDMA appears to cause permanent brain damage, leading to depression, anxiety, and cognitive deficits. The dysfunction produced when oxytocin is "stuck" on or off shows how in healthy people oxytocin maintains a balance between appropriate levels of trust and distrust of strangers.

## Too Rational

Depending on the degree of impairment, people with oxytocin deficiency have varying degrees of cognitive control that can be used as a counterweight. Like high-functioning autistics, or like our friend Stephanie, the highly intelligent survivor of an abusive childhood can train herself to say all the right things in order to appear appropriately empathic. Then again, we can also train ourselves to become moral zombies.

A few years ago, a psychiatrist named Dr. Ansar Haroun called me up out of the blue and asked me to attend rounds with him at the psychiatry clinic he runs inside the Superior Court of San Diego. The specific question that prompted Dr. Haroun to call was the influence of cognition versus impulsivity when it comes to anti-social behavior.

"If people are rational," Dr. Haroun asked me, "why don't they respond to the signals the justice system is giving them through punishment?"

Haroun's view was that psychiatry is more art than science because diagnoses are made based on the clinician's observations— which can never be completely objective—rather than on hard data. He asked me if using economics might provide him with some hard data on these prisoners' decision dysfunctions. It was a great question, so I designed some experiments for prisoners.

One of the first detainees I met had stabbed her roommate twenty-one times. "She bugged me" was her assessment, a reasonable rationale for swatting a fly, but not for killing another human being.

Another was Jenn, a homeless forty-seven-year-old mother of two, who now wore an orange jumpsuit with hands and legs shackled. She was a low-level meth dealer, and her interview was to determine if she would serve eighteen months of hard time or eighteen months in a locked drug treatment center. Jenn's mother had been a drug user who introduced her to meth when she was thirteen so that the mother would have someone to use with. Jenn's life went downhill from there: repeated rapes by her stepfather, running away from home, marriage to another drug user who'd beaten her unconscious. Eventually, Jenn lost her children and was reduced to life on the street. "When my mother calls me in prison and says 'I love you,'" she told me, "I can't say it back."

In the $10 Ultimatum Game, when I asked Jenn how much she'd offer to a stranger, she instantly said $5. I dutifully wrote this down and then asked her how come she was so fair-minded. "When you're a drug dealer," she said, "you cheat, you die." When I asked her for the smallest amount she'd accept as a B-player in the game she said, "One cent." Why? "Easy. I'm homeless."

We found that nearly all but the most highly impulsive prisoners maintained some basic sense of fairness and reciprocity—at least when it came to decisions in the clinic involving money. These studies not only showed the influence of different, rational perspectives on behavior, but also demonstrated that we can cognitively turn down the dial on our empathic response when we need to.

To explore this idea further, philosopher and cognitive scientist William Casebeer and I asked student participants to respond to thirty different moral dilemmas, some of which are personally engaging and some of which allow the participant to remain detached.

We worked with eighty-one people, forty-one of whom were infused with oxytocin; forty were given a placebo. We asked each of them the same thirty hypothetical questions that had been developed by Harvard philosopher Josh Greene for just this kind of experiment. An example of one of the impersonal moral dilemmas was this:

> A trolley car is racing down the track out of control. If you could, would you flip the switch to divert the car to save the five people on board, even if it meant that the car would hit and kill a single individual standing nearby?

To make the same situation more personal, the conditions are slightly different:

> A trolley car is racing down the track out of control. A huge, heavy set man is standing on a bridge over the tracks. If you could, would you throw this one man down onto the tracks, killing him but stopping the trolley, in order to save the five people on board?

In most experiments, people respond differently to the two situations, because one demands that the participant personally engage in deliberately killing one person in order to "serve the greater good" of saving five lives. What we were looking for was the role of empathy in these decisions. What we discovered was that oxytocin infusion had no effect, because both these dilemmas remained hypothetical. Even though the one situation was a little more personalized, it was not "real" enough for the HOME system to play a role.

Which is actually a good thing. There are times when we want individuals to engage in critical thinking and address questions cognitively, without strong emotions kicking in. The problem occurs when we become so good at distancing ourselves that we turn down empathy to zero. Which may be what F. Scott Fitzgerald was referring to when he wrote about the "vast carelessness of the rich." He just as easily could have referred to the vast indifference (diminished empathy) of any high-status group, whether it's CEOs or well-regarded intellectuals.

We ask judges and juries to make rational, disinterested judgments not unduly influenced by emotional appeals. We also want people in high-stress positions such as physicians and first responders to be able to modulate between empathy and the dispassionate state of mind that allows them to do their work. They can't perform an emergency tracheotomy or plan a proper rescue if they're shrieking in horror at the sight of all that suffering.

Once, after I spoke to a group of lawyers, a federal judge came up and confessed that he had no empathy, which led to a poor relationship with his wife and kids. Ironically, his job involved listening to appeals in death penalty cases—which may have been the perfect job for him. At his level of judicial proceedings, the issue is not "can this person be rehabilitated" or "does this person deserve to live" or any such humane consideration, but simply a cognitive assessment of "did this person receive a fair trial under the Constitution."

We want an impartial judiciary, but then again, when *dispassionate* means exclusive reliance on intellect, a Mr. Spock or HAL-like moral obtuseness can enter in. Which is why we still try most cases before judges and juries rather than in front of computers, and why having a jury "of your peers" is so important. Rational, yes, but also human.

As we've already noted, abstract ideas—whether racist notions from eugenics or something as seemingly bland as "rational self-interest"—can impair empathy, and with it moral judgment. Economist Robert Frank at Cornell has shown that undergraduates majoring in economics, where the idea of "self-interest" is central to the discipline, become less trusting and generous in experiments as they move from freshman to senior year. No other academic major appears to have this—for want of a better word—anti-social effect on student behavior.

Which takes us back to Smith's original notion of self-interest and how this idea, distorted and then woven into the fabric of economic studies—and thereby into much of our business culture—has helped create winner-take-all attitudes that do not contribute to long-term prosperity or societal well-being.

## What Psychopaths Do

At the ultimate extreme of ODD is the empathy impairment known as psychopathology. Psychopaths have far more wrong with them than thinking too much or turning people into abstractions, but they are often remarkable for their keen intelligence, and often for their highly effective—but entirely contrived—social charm.

Hans Reiser was very smart, though not necessarily very charming. As a major figure in the socially minded Linux programming community that develops free and open-source software, Reiser was not the type you'd expect to brutally murder his wife. Then again, we've grown accustomed to seeing the neighbors express shock when the family man next door turns out to be a moral monster. In fact, the only thing truly surprising about Reiser's case was the audacity of his request for an appeal. Citing my research, he claimed that his attorney suffered from "oxytocin excess" and thus was insufficiently empathic to provide an adequate defense.

Psychopaths are the opposite of Williams syndrome patients who have great interest in others but little competence in dealing with them. Psychopaths can have incredible social competence on the cognitive level—the trouble is that they simply don't care about anyone but themselves. Their lack of empathy allows them to treat others as objects, and their cognitive skill enables them to get away with it.

Psychologist Anna Salter tells a story about a devoutly religious prison guard—we'll call him Joe—who took pity on a convicted rapist that others had dismissed as a psychopath.

When the prisoner swore he'd found Jesus and changed his ways, Joe spoke on his behalf before the parole board, even promised to take the man into his home if he were released. A few months later, when the newly freed convict raped and murdered Joe's daughter, the guard was of course stricken with grief, but he also expressed a special dismay.

"How could you do this to us?" he asked the man. "We trusted you."

The murderer laughed. "You just don't get it, do you? I'm a *psychopath*! This is what we do."

Joe's willingness to trust made him—and, unfortunately, his daughter—unusually susceptible to being victimized. Distorted ideas from religion, just like distorted ideas from economics, or eugenics, can impair the ability of the Moral Molecule to do its job, which is not so much to make us "good" as to keep us in tune with our immediate environment in the most adaptive way, which usually—but not always—means behaving pro-socially.

Deeply religious people sometimes work so hard to see the good in others, and to be attuned to the needs of others, that they fail to see the warning signs that the person they're dealing with is up to no good—which is not adaptive.

Religion is an indisputably powerful and complex force in human behavior, with a checkered history to say the least. On balance, then, is religion a good thing or a bad thing when it comes to good behavior? Does it enhance the Moral Molecule's function, or is it simply one more impediment?

**CHAPTER 6**

# Where Sex Touches Religion

*Stepping Outside the Self*

An '82 Honda Civic seems a pretty unlikely place for a sudden manifestation of the divine. When Moses heard from the burning bush, it was on top of Mount Sinai. Saul of Tarsus saw the light on the Damascus Road and became Saint Paul. Mystics tend to hide away in caves or ivy-covered priories to commune with God. My overwhelming religious experience—what my mother the nun would call an *epiphany*—happened as I was getting into my car one morning, heading to the library at San Diego State.

I was still an undergrad at the time, living in a sketchy part of town in a Spanish-style apartment building that had been officers' quarters in World War II. The owner was a guy known as Dr. Dean who performed hypnotism in nightclubs around town, and most of the other tenants, who were well into their eighties, seemed to have been there since the Japanese surrender. When I came home at night I'd pass the hookers on the street corners just a couple of blocks away. One night I remember a girl standing there who must have been six months pregnant.

Truth be told, I think the wrong-side-of-town ambience may have added to the experience, the same way you might appreciate a wild rose even more if you found it growing up through a crack in the asphalt. I also think I'd been preparing myself for some kind of breakthrough for quite a while.

I hadn't gone to college right after high school. I'd been bored as a student, and even though my father worked at the University of California Santa Barbara and would have qualified for a discount if his kids went there, we'd never actually discussed further education. The way my sister sized it up, my parents viewed us as souls to be saved rather than as future adults to be nurtured along. Which no doubt intensified my desire to distance myself from their approach to religion.

I moved out right after I graduated, and I got a job selling shoes. Within a year I was managing two shoe stores, bringing in a paycheck equivalent to about $80,000 a year in today's dollars. But finding meaning and purpose was not part of the benefits package. Despite my recoil from organized religion, all those years of bells and smells as an altar boy had put down the roots of a spiritual seeker. I just needed to find my own answers my own way.

Eventually I went back to school, and along with lots of math and biology courses I immersed myself in philosophy and the history of religion. Most recently I'd been studying Confucian thought and the idea of the *wu wei*—the Daoist concept that there is a natural rhythm to the energy of life, and that the task is not to master or disrupt that rhythm but simply to embrace it. Once you're in sync with the flow, it carries you along and the chances for a happy and harmonious life increase exponentially.

These were still pretty abstract notions for me when I got into my Honda on the morning in question. This was late in the fall, and there was a cool Pacific mist still hovering over the streets. The night before, I'd left the radio on—tuned to National Public Radio, as it turned out—but when I cranked the engine, what I got was not *Morning Edition* or some interview show, but a sudden blast of music, Pachelbel's Canon in D. I knew the piece, but the version I heard on this morning seemed to be moving at a faster tempo. Years later I actually tracked down the particular recording and learned that it was performed by Sir Neville Marriner and the Academy of St. Martin in the Fields.

Thirty years have passed since that chilly morning, and that piece of High Baroque has become a cliché of wedding ceremonies and high school graduations. But for me, then and there, listening to it was utterly transformative. The three violins intertwining over the endlessly repeating bass line. Variation after variation on the same beautiful chords and melody, piling on top of one another, building in intensity until the interwoven rhythms and themes seemed to grab me and pull me out of the literalness of my ratty car parked at the curb in my ratty neighborhood. As I listened to the music, my entire body pulsed with an overwhelming sense of love and belonging and peace. Tears streaming down my face, I felt a revelatory sense of connection with the entire universe. Every living thing, every inanimate molecule swirling in every galaxy, seemed woven together into a single warm embrace. I was floating in a boundless sea of love, with waves of kindness and connection washing over me.

And then the moment passed.

Okay—I wasn't on drugs. So what was it? One more stressed-out, sleep-deprived student on the verge of a nervous breakdown?

Or maybe it was some sort of seizure. A mini-stroke, perhaps? Or, in my mother's terms, was God trying to speak to me?

I'm a scientist, and people like me are supposed to be intensely secular. In just the past few years, several "people like me"—by which I mean neuroscientist Sam Harris, philosopher Daniel Dennett, and evolutionary biologist Richard Dawkins—have, in fact, written hugely successful books that trash religion in the harshest terms.

Of course, religion makes a pretty easy target when you focus on the trail of discord and bloodshed it's dragged across the pages of history. But I think the hostility reflected in these books, as well as the popularity reflected in their bestseller status, is more a reaction to the way personal religious beliefs have intruded into public life. Secularists object to the unspoken requirement that our political leaders have to say all the right pious things and make all the right pious gestures, like attending prayer breakfasts. And it's especially annoying for non-believers when believers act as if everyone has to be among the faithful in order to be morally grounded.

But while scientists like Dawkins heap scorn, other scientists compile data showing that religion can be, in fact, very good for us. Solid studies demonstrate that, on balance, religious people are happier than others, and that weekly attendance at a religious service, if only through social effects that improve the immune response and resilience to stress, makes people demonstrably healthier. In addition, there's no doubt that religion has contributed mightily to human betterment through charity hospitals, suicide hotlines, schools, food distribution, and, above all, moral guidance.

As for some sort of secular revolution ever taking place in our culture, the evidence strongly suggests that the religious impulse

133

has been around way too long and is far too deeply ingrained for us to think that it's going to go away anytime soon. Anthropologist Lionel Tiger estimates that 80 percent of the human race is affiliated with some form of faith-based community. But then again there are more than four thousand varieties, each with its own way of finding God, or worshipping God, or trying to benefit from interactions with God. Which brings us back to that long list of cases in which religion was the source of discord and violence. The fact is that religion can bring out the very best in people as well as the very worst. Some say that religion is essential for morality; others, that it divides more than it unites. On balance, then, does religion contribute to the good stuff we see emanating from the Moral Molecule or not? I use experiments to explore the world, and experiments begin with a hypothesis. Everything I know about biology tells me that nature is conservative, that it uses the same systems for multiple purposes, so a good assumption to start with is that the Moral Molecule that connects us to others also facilitates what many perceive as connection to God. As we've noted many times, we are, after all, an obligatorily gregarious species.

Another premise to start with is that every religion, one way or another, seeks to achieve something akin to what I felt in my Honda on the way to the library that morning in San Diego. The Greeks called such heightened experience *ekstasis,* or "stepping outside," as in "stepping outside the self." It's when we move beyond the boundaries of the self that we're able to connect with something bigger—the essence of the religious quest.

I wondered why this desire to reach transcendent states was so nearly universal. I also wondered how oxytocin was involved. To find out, I'd have to induce something akin to the religious experience in my lab.

My first thought was to bring in a recording of James Earl Jones narrating the Bible, but that didn't work. Too Darth Vader. So then I considered other clips of religious people speaking about their faith, including the Dalai Lama and Father Thomas Keating, a Trappist monk and Christian mystic of the Thomas Merton variety. The overriding problem, however, was that most college students, including those raised in a religious tradition, are not very religious, so I was fighting a losing battle trying to tap into the conventional imagery. So I thought about uplifting art, including religious art, and expansive nature scenes with and without music. At the other extreme, I considered using a sensory deprivation helmet that could induce hallucinations.

Sometimes religious devotees attempt to step outside themselves by going deeply inward through meditation and prayer. When scientists scan the brains of intense meditators, whether Franciscan nuns or Buddhist monks, they find that among these devotees, the parietal lobe—a part of brain that helps maintain the sense of self—substantially decreases activity. Which sounds like a pretty good way of achieving the *ekstasis* that enables the self to feel that it is merging with the universe.

I first dipped my toe into investigating religion with a study in which we taught students either standard mindfulness meditation or a meditation practice known as metta. That is a Pali word that translates as "compassionate love." After four weeks of training, both groups showed increases in trust, generosity, and compassion, but the metta group had the largest increases. In the Trust Game, the metta group showed a 33 percent increase in trust, while the mindfulness meditators increased only 7 percent. We also imaged the brains of those in both groups while they meditated and while they made decisions that involved sharing money with other people. We found that the executive function areas of

135

their brains quieted down, shifting the focus away from self. The HOME circuit also lit up during social tasks, showing clearly that oxytocin was driving their decisions toward greater compassion.

I also considered tapping into older traditions that approached *ekstasis* by way of a direct assault on neurotransmitters. At the Oracle at Delphi, the priestesses of Apollo stepped outside themselves by way of trance states in order to divine the future. Recent studies suggest that they were very likely helped along by breathing the brain-scrambling fumes of ethylene gas seeping out of the rock beneath their temple. In the American Southwest, the faithful step outside themselves and into ritual space and ritual time by taking peyote. In the Andes, the religious tradition includes a hallucinogenic tea called ayahuasca. Flower children never achieved much respect as a religion, but they were making the same effort to step outside themselves and to find peace and love and enlightenment with LSD and psilocybin. Today the drug of choice gets right to the point with a name that's the English translation of *ekstasis*. (Again, as mentioned in the last chapter, this drug, ecstasy, also produces significant amounts of brain damage.)

Christian mystics, thinking more in terms of asceticism than biology, fasted in order to induce the *ekstasis* of "visions." Of course, the clinical term for those experiences is "hallucinations." In the Middle Ages, the visual hallucinations associated with mystical devotion seemed more prevalent than now, but so was eating bread mold, particularly a strain called ergot that infests rye, wheat, and barley. Ergotamine, a compound within the ergot bread mold, is one of the components used to synthesize LSD.

It's also an accepted part of religious traditions that great visionaries such as Moses, Muhammad, Saint Paul, and Joan of Arc were "visited" by the angels, the gods, or God. Turns out that the descriptions of these visitations match up pretty well with descriptions of seizures caused by temporal lobe epilepsy. I've seen many temporal lobe epileptics in clinic and they are typically hyper-religious, always wanting to convert me. Other illnesses—schizophrenia in particular—can induce particularly religious compulsions, including the claim of being God. If you ever saw the film *The People vs. Larry Flynt,* you know that Flynt, the porn king, went through a phase during which he wanted to turn his flagship publication, *Hustler,* into the world's first Christian skin magazine. His editors thought he was nuts, and he now certainly agrees. "I became deeply religious," he told a friend of mine. "I sought professional help, and I was cured."

The other most common route to *ekstasis* is through music and dance, which often get very lively, and is where we get our word *ecstatic.* Thousands of years ago, organized religion was born when our ancestors discovered that there was a multiplier effect to this particular approach. Think of tribal dancing around a fire in some village in the rain forest, the whirling dervishes of Sufi mysticism, or the chanting and singing, speaking in tongues, and generalized swaying and hand-clapping of charismatic religions like Pentecostalism. When music and dance are part of a communal ritual, it takes on even more power because it allows individuals to feel connected with God while also connecting with one another. So it would appear that they're doubling up on oxytocin and the serotonin that flows from it.

Dance can express joy or sorrow, but done right, it always leads to getting in touch with life on a deeper level. In *Zorba the Greek,* an uptight Englishman visits an island in the Aegean, experiences several emotional traumas, then has his last big hope for economic success come crashing down around his head. When that moment of complete catastrophe arrives, he doesn't cry or moan or curse

God. Instead, he turns to his earthy and elemental guide to village life on a Greek island and says, "Zorba, teach me to dance."

Dancing is just a very human thing to do. In case you've never noticed, children will dance spontaneously. My mother tells me that when she was a novice in the convent, the nuns would square dance with each other. She describes how moving around and laughing with the women she didn't know well or had never liked much broke down the barriers and made her feel much closer to them.

Dance seemed like a particularly good subject for experimentation because it also allowed me the option of studying ritual without religion. After a brief survey of styles, I settled on an old New England variation on the square dance called contra dancing. Its experimental virtues included the fact that everybody does exactly the same thing, and that each person partners intermittently with everyone else. It also so happens that I found a group of contra dancers not far from where I live who were willing to shed their blood for science.

## Dancing with the Scientists

My grad students and I showed up at the Woman's Club of South Pasadena at six o'clock on a Saturday night. The dance wouldn't begin for two hours, but we needed to set up our blood draw tables, rice paper screens for privacy, centrifuge, tubes, and pipettes, all pre-labeled and organized in exquisite detail. We got everything ready with plenty of time left over to go out for sushi, and when we returned the doors were open and the band—which I'd paid for in return for everyone's gracious participation—was setting up.

The people who came to Pasadena to swing one another around the dance floor ranged from veterans in their sixties and

seventies to one very cute college freshman who'd only participated once before. A woman in her sixties told me she came
because she had awful back pain every day, but when she joined
the dance the pain disappeared. I asked her if she thought it was
the exercise or the people that served as an analgesic. She thought
for a moment and said, "Both."

Two-thirds of the fifty attendees that night agreed to be our
guinea pigs. We took their blood before the dancing began, and
then again after either the third or fourth dance. We'd have to wait
for lab results to offer any conclusions about hormonal change, but
as for social effects, these were obviously very happy people, ooz-
ing with connection, even without any particular recourse to God
or godliness. After we finished I came down during a band break
to thank the volunteers, and the entire group stood up and cheered.
I've done experiments that got lots of media attention, but this is
the only one that ever earned me a standing ovation.

When the results came back from the lab, we found that, on
average, oxytocin rose by 11 percent across the range of age and
gender. We were pleased but not surprised, and then the results
got even more interesting. Before and after the dancing, we also
showed our volunteers diagrams of the social setting and asked
them to put a little $x$ where they saw themselves fitting in. The
higher the individual's OT release, the nearer to the center of the
group that person placed the mark. The average increase in closeness to others after dancing was 10 percent. But the result I found
most striking was that, after an hour or two of contra dancing,
even these generally very secular people had an average increase
of 3 percent in the extent to which they described themselves as
being "closer to something bigger than themselves."

With these results in hand, I next wanted to reverse fields and
study religion without ritual, so I turned to the Society of Friends,

also known as the Quakers. Their religious services don't involve singing, dancing, or any empathy-inducing imagery like the crucifixes and stained-glass windows that I was used to. Their services don't even have preaching. Instead they gather together to focus communally on each individual's contemplation of—and presumably connection to—God. Before running an experiment I attended a service and found it very relaxing to sit for an hour in community and meditate.

Did oxytocin play a role in any of this? To find out, I gained permission from the Religious Society of Friends in Claremont to set up my blood tubes and needles, tourniquets, and ice in a conference room where you might expect to find coffee for the after-church fellowship hour. Seventeen brave souls gave us two tubes of blood before their one-hour communal meditation, and returned to give us two tubes afterward.

Without ritual, there was no overall change in the average OT. But this was because roughly half the group had a powerful increase, and the other half had a powerful decrease. It seems that, while sitting in silent contemplation can create a feeling of increased closeness in some, in others it creates the inattention known as boredom. But at least there was an overall drop of 7.3 percent in the stress hormone ACTH. After meditation, participants reported feeling 7 percent closer to members of their society, and 4 percent closer to "something bigger than themselves." Even religion without ritual does something.

## Making Meaning

So meditation can calm us, and turn us away from self-interested concerns, and ritual—even dancing—can rev up the oxytocin that will make us feel more connected to others as well as to something larger. But how do we get from there to God? And do we need God to make us moral?

As we've seen in other contexts, the human brain is an instrument for making meaning. In the Heider and Simmel film from the 1940s that I've already mentioned, people presented with three moving geometric forms can come up with a drama about good and evil, victims and victimizers. So it's not too hard to see how warm feelings of connection, even colors and shapes and sounds created by experiences with people or simply by anomalies in the brain, can be swept up into religious narratives of a Creator and First Principle in charge of all this. Most theories of dreaming converge on the idea that what's happening is a processing of random bits of information during sleep, and that the often bizarre narrative that we remember—the dream—is the brain's attempt to weave all these random bits into a coherent narrative. Expanding that perspective to the entire human race, the Jungian school of psychology considers religion to be a collective dream shared by vast numbers of people.

Even before our ancestors had pulled together any explanatory story lines, they must have been painfully aware that nature was a lot more powerful than they were, which led to feelings of fear and awe. Consciousness and the meaning-making that came with it, led to an effort not just to make sense of those same fearsome and awe-inspiring natural forces, but to keep those forces appeased. When any natural force makes things happen—as when lightning strikes a tree and causes a fire—our hyper-social, meaning-making brain can assign human intentions to that power, a mental habit called anthropomorphism. Social neuroscientist John Cacioppo has assessed individuals' degrees of loneliness, then shown them photographs of distant objects in space such as the Horsehead Nebula, swirling and moving through the darkness of infinity. The lonelier the people, the more they are likely to anthropomorphize these vast clusters of stars and gases, not only giving them personal characteristics but attributing human intentions to them as well.

In similar fashion, ancient humans observing nature must have recognized the positive influence of the sun and the spring rain, along with the destructive influence of storms and drought and lightning. As products of natural selection, consciousness and the narratives it concocted were very much concerned with the same effort that occupied most of our other energies, which was the effort to keep life going.

So it should come as no surprise that, along with stone tools, the oldest human artifacts ever found are religious totems in the forms of voluptuous female figurines, symbols of fecundity and the mysteries of reproduction. In some cultures the phallus was also venerated, and both these sexual symbols were employed in devotions meant to appease some higher power, thus ensuring that the cycles of life would continue, that the rains would come, the delta would flood, and that the crops and the game would be plentiful.

Tens of thousands of years later, even after the advent of armies and city-states, mathematics and philosophy, poetry and sculpture, the ancient Greeks still worshipped the life force they called Eros, also known as sex. In the same way that oxytocin and testosterone operate as antagonists, the Greek myths held that Eros, the god of sex, was the child of Aphrodite, who represented love, and Ares, the god of war.

But Eros was also a mainline to *ekstasis,* and to the release of oxytocin that surges at the moment of sexual climax. Another direct approach to *ekstasis* was the worship of Dionysius, the god of epiphany and all other things wild and irrational. It was the ecstatic rites of Dionysius that gave rise to Greek tragedy, which offered a special form of stepping outside called catharsis, in which the members of the audience empathized profoundly with the characters on stage, recognizing and absorbing the pathos of our common humanity.

The reverence for reproductive power and sexual ecstasy that began with sensual-looking female figurines and phallic symbols, and that led to fertility rites and ecstatic dance, eventually carried over into the Christian world. No matter how much the church tried to suppress sex, the erotic was never that far removed from the spiritual as one more way of stepping outside. You can see the power of Christian *ekstasis* in the Bernini statue that sits in the church of Santa Maria della Vittoria in Rome. The subject is the seventeenth-century Spanish mystic Saint Teresa of Avila, and the representation of her face captures what she called "the devotion of union" with God, a union that, judged by her expression, appears absolutely orgasmic. "It is love alone that gives worth to all things," she said. "God is Love" is, of course, a Christian mantra, from papal encyclicals to the bulletin boards in Protestant Sunday schools.

Christianity actually makes much of four different types of love as represented by four different Greek words: *eros* for erotic love, *storge* for parent-child love, *philia* for brotherly love, and *agape* for love of God. But the distinctions still get blurry. For instance, a Baptist hymn called "In the Garden" makes faith sound an awful lot like a love affair with Jesus. In the song, the speaker comes to a garden alone, "while the dew is still on the roses." The refrain says:

*And He walks with me, and He talks with me,*
*And He tells me I am His own;*

*And the joy we share as we tarry there,*
*None other has ever known.*

The second verse begins:

*He speaks, and the sound of His voice,*
*Is so sweet the birds hush their singing;*

Update the language a bit and that gushiness is not so far removed from pop lyrics to songs like "He's So Fine" or "My Girl."

A hymn like "The Old Rugged Cross," with its imagery of incredible love and incredible suffering, is certainly as oxytocin-inducing—and empathy-inducing—as our video about Ben, the little boy with cancer. But the classic "altar call" hymn in Baptist churches, "Just As I Am," ends every verse by repeating a line with overtones that are hard to ignore:

*Just as I am, without one plea,*
*but that thy blood was shed for me,*
*and that thou bidst me come to thee,*
*O Lamb of God, I come, I come.*

That's not just empathy—that's *ekstasis*.

The most obvious element of Eros and the ancient fertility cults to be transposed onto Christianity is, of course, veneration of the Madonna. Christianity being Christianity, there was the requirement that the powerful female force had to be chaste, but this dovetailed with another ancient concept fundamental to many religions—virgin birth. In ancient Egypt, Isis was supposedly born of a virgin, as was the Babylonian god Marduk, and the Hindu god/man Krishna. In Persian mythology, Zoroaster's mother was supposedly impregnated by a shaft of light. There were similar origins proposed for the Greek god Perseus, and even for Roman emperors after they managed to get themselves deified.

Virgin birth was a way of establishing a union between God and humanity, not a surprising objective for such a highly social species. This connection was all the more appealing for taking place down here on our level, within the realm of hormones and neurotransmitters, within the realm of biology. In other words, the cosmos works the same way a human family does—with love and caring.

Even today, studies show an explicit link between the religious impulse and the literal urge to reproduce. Young singles were exposed to large numbers of attractive people of their own gender, then, presto, the percentage of those singles who professed religious feelings increased. The reason why? It's adaptive. In an environment where the sexual competition is more intense, a free and easy attitude toward sex is not the best way to land a reliable mate. So individuals opt for the kind of conservative, monogamous behavior associated with most religious teachings. Thus it's not just that people learn a conservative sexual morality through religion, but that conservative attitudes about sex and family life *cause* people to embrace a religious lifestyle as a way of attracting and retaining a high-quality mate.

Earlier traditions used sex itself as a way of achieving *ekstasis,* and the temple prostitute who could bring worshippers to a state of grace through union is a common feature of most of the world's ancient religions. But even today, when physical contact goes no farther than the handshake after communion, connecting with other people while also connecting with the divine seems to be the center of the target for religious ritual.

## *Ekstasis* Goes Public

As we saw with our contra dancers, when stepping outside the self leads to greater connection to others, oxytocin release leads to stress relief and soothing of the nerves through the HOME system. So religious ritual provides a positive, physical outcome even before it introduces reassuring concepts like life after death, eternal rewards for good behavior, and reuniting with loved ones. Oxytocin release stimulates the release of serotonin to reduce anxiety and feel calmed, and dopamine makes it "sticky," which is to say, something you'll want to keep doing.

Oxytocin induces empathy, which creates compassion, which helps groups hang together with common purpose. It also reinforces trust. In the book of Hebrews, Paul writes about "trust in things not seen." The Greek word Saint Paul used to describe Abraham's relationship with God, *pistis,* is often translated as "faith." But in Greek mythology, Pistis was actually one of the spirits that escaped from Pandora's box and flew away to heaven. It was the spirit of trust.

When oxytocin release, empathy, and connection with God came through public rituals, it provided connection with other people who were, in the moment, sharing the same feeling of connection with God, which was a double whammy of *ekstasis*—like a cosmic pep rally. This was just the kind of thing that could inspire a group to hang together and overcome and endure. Which is why most religious practice is communal.

Darwin argued that religious beliefs arose and endured because they made societies more willing to cooperate and to sacrifice for the common good, which enabled them to outcompete groups of self-centered individuals who lacked the social glue of a shared faith and a sense of purpose beyond the self.

But this double whammy is also where we find the downside of revved up religious fervor. The same in-group connection that creates such tremendous empathy, which creates the willingness to sacrifice for the common good, can also help fuel hostility to any out-group. When you're so revved up on the ecstatic feeling that you have God on your side, members of other groups become not just "the other." They can become "sinners" or "heathens" or "children of the devil" that need to be wiped out. The imagery of *ekstasis* has proved startlingly effective in the stagecraft of the Ku Klux Klan with their burning crosses and white hoods, and in the huge rallies that Joseph Goebbels orchestrated for Hitler during the Nazi era.

I wanted to understand why some religions produced in-group biases, but I first needed a baseline of a strongly bonded secular group for comparison. After several meetings and participating in a grueling patrol-ambush exercise in the mountains that left me bruised and bleeding, I was able to convince Claremont's ROTC battalion chief Lieutenant Colonel Bill Fitch to permit his cadets to play the Trust Game, both with other cadets and with non-military volunteers from the student body at Claremont. To strengthen their in-group feeling, the cadets made decisions immediately after they'd engaged in fifteen minutes of marching outside my lab, a typical "ritualized behavior" for them.

I also had a group of students who self-identified as evangelical Christians play the Trust Game among themselves, and again with non-evangelical students. These young people worshipped and sang in my lab for fifteen minutes for their ritual. To serve as controls, we recruited a bunch of non-affiliated students and divided them into "Reds" and "Blues" to form completely arbitrary in- and out-groups. Control participants played the game of telephone for fifteen minutes with their own color to make their group salient. I also took blood from everyone before and after their rituals.

Here's what we found. Among the controls, the B-players—those who might return money after the experience of being trusted—returned 23 percent whether they were dealing with an in-group member or with someone from the out-group. In other words, the artificial and arbitrary tribal distinctions we'd set up for them—Red and Blue—made no difference.

Among the B-players who were cadets, there was a significant in-group bias in how much they returned—51 percent if you were "one of us," versus 40 percent if you weren't. The same was true for the evangelicals. Their bias was about the same—38 percent returned to those from within their group of fellow Christians, and 28 percent to those who were outside the fold.

It was among the A-players—those who made the initial decision based on whether or not they thought they could trust a B-player—that the evangelicals stood out for having a much stronger in-group bias than even the revved-up ROTC members. Evangelical A-players transferred 84 percent of maximum to in-group B-players versus 61 percent to out-group B-players—a spread of 23 percentage points. Among the ROTC cadets, the in or out distinction was 81 percent of maximum versus 74 percent—only 7 percentage points. Among the controls, the transfer was the same no matter whether the B in question was a Red or a Blue. On average, they transferred 58 percent of the maximum amount available—an indication of far lower levels of trust without some kind of meaningful group affiliation.

Part of the difference in behavior by evangelical participants was due to their higher stress levels: 28 percent greater than controls. They were simply more anxious about interacting socially (and ROTC cadets were 17 percent less stressed than controls).

Benefiting one's own before strangers is to be expected. But the problem with an in-group bias, especially one hopped up with ritual, is that it limits the opportunities for connection that we crave as social creatures. The messages of the world's major religions resonate with us today because they preach the promise of universal connection and love. The need to belong is part of our human nature, and religious figures like Jesus and the Buddha seem to have figured out how to love anyone and everyone and to have love come back to them in return. But it would appear that in-group bias makes it all too easy to love some a lot more than others. Limiting the love to your own group actually imposes an economic penalty. Because they were so much less trusting of outsiders, the evangelicals took home 9 percent less from their Trust Game experience than the ROTC cadets.

Throughout history, whether in a socially useful way or in a diabolical way that leads to genocide, groups have tried to invoke a higher power to get people to follow the rules. We've already discussed the societal importance of a willingness to punish, but not everyone is willing to punish every time, because playing the heavy can carry costs to the punisher. It's awkward, stressful, and you always have to worry about straining a relationship or inviting retribution. It's also true that (at least before Google Maps came along) real life doesn't allow for watchful eyes to be on hand everywhere, and at every moment, to monitor everything each of us does.

Having an all-knowing, all-powerful God, however, allowed societies to outsource the punishment. Being able to say, "It's God chastising you . . . not just me," takes some of the pressure off everyone else who, despite the *urge* to punish, mostly wants to mind his own business and get through the day. Putting God on the case also delegates the punishment *upward*. This all-knowing

God is not just a little frightening, and the fear factor helps to strengthen and internalize the do-right message because, ultimately, there's no place to hide. Because God is everywhere and all-knowing, ultimately there will be consequences for sin, even if you appear to be getting away with murder right now.

Studies have shown that any sort of priming with the sense of being watched can induce better behavior. The display of the Ten Commandments, a pair of eyes on a computer terminal, telling children that a magical Princess Alice is watching their game, or telling students about the presence of a former graduate student, now deceased, whose spirit haunts the lab—any of these devices prompts people to behave more pro-socially and less selfishly.

Combining the idea of a supernatural surveillance system with a form of payback system—karma, the wages of sin—allows other group members some of the pleasure of punishing. Just imagine the torment to come when that scofflaw gets what's coming to him!

## Magical Me

When social scientists ask people to describe God, it turns out that there is very little consistency in the attributes they come up with. The descriptions don't all include an older white male with a beard, a robe, and sandals, or a big loving mama floating in the clouds, or a giant computer in the sky. In fact, the only factor that emerges from an analysis of individuals' concept of God is that for each of us, God appears to be a projection of the self—me and my attitudes, wants, and wishes—albeit a "me" with exceptional powers.

It makes sense, then, that our concept for universal moral guidance would map onto the same physiological mechanism that modulates the two sides of our own, individual moral behavior. In each of us there is oxytocin to nudge our behavior toward love and connection, but also enough testosterone to activate fear and punishment. So it is with the Big Guy (or Loving Mother) watching over us. God, the Ultimate Moral Judge and Enforcer, aligns with the influence of testosterone. God, the Ultimate Source of Bonding and Love and Concern, aligns with oxytocin.

Any wonder, then, that belief in God can inspire acts of both tremendous compassion and vicious, sectarian violence?

The first, struggling groups of hominids, physically outclassed by chimps and no match for lions or packs of wild dogs, needed to bond together, get along, and help keep one another alive. Accordingly, the animistic gods of primitive hunter-gatherers and ancient fertility rites were driven more by oxytocin. As tribes became larger and more genetically diverse—in other words, not everybody was family—survival required greater degrees of rule enforcement, so God picked up more testosterone. Nomadic tribes were more likely to encounter other groups, which created more of an us-versus-them dynamic, which increased the need for God to be partisan as well as punishing—no longer an Earth Mother but "the God of High Places" who could be called upon to smite the other guys. Certainly the God of the Old Testament is nothing if not a strong father figure, and not a very nice one at that. He's always saying things like "I am a jealous God" and "I am an angry God." And as the Bible tells us again and again, this God of Wrath, the ultimate bad boy, was ready to summon floods and to destroy whole cities at the drop of a loincloth.

Small-scale societies could rely at first on the natural give-and-take of human nature. Our species led with generosity and trust, but with retribution—or at least the withholding of acceptance and generosity—whenever there were violations of trust. Then God was invented to reinforce these pro-social tendencies and to back up punishment of anti-social tendencies. Eventually the rules said to come from God were codified and given the force of secular law. In time these laws became fixed, as in the Ten Commandments and the Code of Hammurabi, which laid out the requirements very clearly, especially for the 5 percent of any population who lacked the oxytocin receptors necessary to bond and behave morally without external enforcement. But even for the other 95 percent, having a bright line that society sets for good versus bad is helpful, especially given that moral sentiments are, of course, fallible and subject to physiologic whipsaws.

Even though "an eye for an eye" characterized most of these ancient forms of jurisprudence, other forces focused on working out non-zero-sum solutions to survival, which created more of a role for compassion. Testosterone-driven warrior gods like Jehovah and Zeus and Jupiter continued to be in charge throughout the Classical era, but the pagan pantheons always made room for a variety of voices, including the ecstatic and the erotic. Meanwhile, the more secular philosophies of Greece and Rome infused moral thinking with a strong element of reasoning.

The Jewish world also had its mystics as well as its philosophers to leaven the harshness of the God of Wrath. These included Rabbi Hillel, who came up with the Golden Rule several decades before it appeared in the New Testament as one of the teachings of Jesus.

Clearly, though, the approach to religion articulated in the New Testament, a perfect amalgam of various trends coming to a head just as the classical world reached the apex of its power and then collapsed, was a powerful new idea at just the right time.

The testosterone tradition of Zeus and Jupiter and Jehovah was maintained through the continued reverence for the ancient Hebrew scriptures, which Christians began to call the Old Testament. The

Greek philosophical tradition was maintained by adopting a view of the universe built on Plato's concept of an alternative, perfect world of Form, a spiritual realm that we ordinary humans see only "through a glass darkly."

Roman power had disrupted the old ways while the Pax Romana brought a variety of cultures into contact with one another, with mobility lessening the ties of religions grounded in a particular place. Christianity filled the void by teaching that the kingdom of God was not inside this or that temple located on this hill or above that holy spring, but that it existed within the heart of each individual believer. In this respect, Christianity actually began as a bottom-up moral force, with dozens of different approaches. But those who wanted a more top-down orthodoxy eventually won out, and the "gentle shepherd" found his legacy co-opted by the ultimate command-and-control organization, the Roman Empire, which morphed into the oh-so-hierarchical Roman Catholic Church.

Even so, the bottom-up idea persisted in what was Christianity's most significant innovation, which was, after thousands of warrior religions, a return to compassion. Just as the cult of Dionysius had huge appeal for women and slaves and others who were denied the privileges of full citizenship in a warrior society, the cult of Jesus offered God's love for everyone, no matter now humble, no matter how despised by the wealthy and powerful. It was the loving and forgiving Christ, the oxytocin-enriched lamb of God, that made the Jesus cult a major spiritual force capable of lasting for more than two thousand years.

In Asia many historical religions had focused on compassion, following precepts that offered a release from suffering by putting a stop to the endless cycles of reproduction and reincarnation. A God of Wrath seemed unnecessary because rule-following was already much more deeply ingrained in these less individualistic, more group-focused Asian societies. Then again, compliance was enforced by the presence of one's ancestors and by a heavy reliance on shaming for bad behavior, especially behavior that brought dishonor upon one's group.

So once again, in religion as in all else, the underlying truth of the Moral Molecule is that we are not so much naturally compassionate or aggressive, or generous or ruthless. Rather, we are naturally adaptable. The opposing hormones that regulate us allow us to go either way, depending on the circumstances. It so happens that being generous and kind and trustworthy is, most of the time, by far the best way to go. So we embraced moral exemplars who could guide us on how to do this: Jesus, the Buddha, the Dalai Lama.

Like religious imagery, rituals of communal eating and laying on of hands have always been a part of religious fellowship because they boost oxytocin. In our Trust Game studies, those who showed the biggest spike in oxytocin and who were most trustworthy were also those who described themselves as being religiously committed. Sorry, Richard Dawkins et al., but these religious people also scored highest in measures of life satisfaction and emotional well-being. The critical factor in making this all work out for the best is the one emphasized by gurus ranging from Jesus to John Lennon: All you need is love.

## Transcendent Love (and Mud)

Which brings me to a much more recent epiphany in my life, a moment in which the rush of oxytocin truly allowed me to step outside.

After two years of preparation, I had finally received permission to go to Papua New Guinea to take blood from tribal warriors before and after they performed a ritual dance. The experiment would test if oxytocin release was universal. Malke village was in the Western Highlands, a rugged, green place of volcanoes and near-constant rain. It was also thirty hours of travel away from California, and when I arrived I have to admit I was a little overwhelmed. In part I was simply knocked over by the smell of all these unwashed bodies. There were about one thousand people, living pretty much as our evolutionary ancestors had. They built thatched huts and survived on yams and cabbage. Truth be told, they were at most two generations removed from cannibalism. The men were so caked in dirt that, once we began our experiment, I had to use four or five alcohol swabs to get down to skin before we could take their blood. For "sanitary facilities" they used a ditch, with no soap and water for cleanup—at best a leaf to wipe their hands. They went barefoot in the mud and wore secondhand clothes.

Reeling from culture shock and serious jet lag, I was also flummoxed by the fact that our liquid nitrogen for cooling the blood samples had not arrived, so I sat down on a hillside to try to figure out what to do. A very tall Westerner with a pale face, I was obviously a curiosity, because, little by little, some of the villagers began to gather around. A few more moments and I'd drawn a crowd, the people ranging from toddlers to toothless grandmothers. They sat down around me and they watched me, gradually drawing closer. The local tradition of "having" guests for dinner did cross my mind, but "In for an inch, in for a mile" I always say. The children were shy, so I began to make goofy faces. They began to laugh, and then everyone smiled. Then everyone began to draw closer still. They all wanted to touch me, to shake my hand. They began to offer me their headdresses made of animal skins, feathers, and grass. I began to flirt with the old ladies—one of whom kept giggling and jabbing me in the ribs—and then I simply breathed in the smell of the jungle and for a moment let go of all my Western thoughts, and biases, and worries. It was a life-changing experience.

These were the friendliest, most joyous people I had ever been around. Because they're able to attain all their necessities with about an hour's labor a day, they sit around socializing most of the time. (When I analyzed their blood, their stress hormones were like those of someone who was barely conscious.) And yet they also could be incredibly industrious and caring. We asked them to build a hut for us to protect our generator and electrical equipment from the constant rain—a tarp over tree-branch poles—and they went to the trouble to decorate each pole with ferns and purple flowers. They were also incredibly generous in other ways. I watched them bludgeon a pig and slowly roast it, pulling off the outer layers of skin as it cooked, then with great ceremony the chief distributed the luxury of meat, an equal share to each family. Everyone waited for all the meat to be distributed before eating. Later, when it was time for us to leave, these people, who had nothing, held a ceremony to give each member of my crew a beautifully wrapped gift with a note from the chief about why this gift was important for each of us to have.

The results confirmed that, like Westerners, isolated farmers in Papua New Guinea do release oxytocin during rituals. But merely by being here, I felt that I had learned something incredibly powerful. I was halfway around the world, surrounded by people who could not have been more different from me. I was a complete stranger to them, and yet they absorbed me into their village almost immediately. We had gotten down to something truly primitive, like the blood of birth or the blood of battle. It was the love and empathy of oxytocin, bridging the gap of thousands of miles and eons of social differences. I was covered in mud but I felt uplifted, just as I'd been so long ago listening to Pachelbel's Canon. I experienced that same sense of love and belonging and peace, of connection with the universe. This was all the religion I'd ever need. I felt truly alive, immersed in this sea of humanity.

# CHAPTER 7

# Moral Markets

*Liquid Trust and Why Greed Isn't Good*

S hortly after our first oxytocin infusion paper landed me on national television, I spotted a new product being sold on the Internet called "Liquid Trust." Only $40 for a two-month supply, what a deal! The Web site cited my research and the media coverage and was loaded with testimonials. Which I found rather odd, especially when I considered all the snorting and eye-watering discomfort required to get an effective dose of oxytocin into the brain, not to mention the fact that the effect lasts for only a few hours. Then there were those pesky FDA regulations on oxytocin inhalers that I knew all too well.

I shrugged it off and went back to the lab, but a few months later a television producer from the now-defunct Fox News morning program *The Morning Show with Mike and Juliet* called to ask me to appear with a spokesperson for the Liquid Trust company. So I looked a little closer at those ads, which read: "Each 1 oz bottle (2 month supply) of Liquid Trust contains the following ingredients: Purified water, SD Alcohol, and Oxytocin."

Oxytocin, as we know, is a prescription drug. So they were either lying about the ingredients or violating federal law by selling it over the counter. But then I noticed the stroke of true sleazy genius. The directions said to *spray it on your clothes!* They weren't dispensing a drug without a prescription—they were selling a very, very expensive air freshener! Duh! If only I'd known how easy it was to create an atmosphere of trust, I could have saved myself and my scientific colleagues all over the world a hell of a lot of trouble.

I went on the television show, and with their spokesperson looking on ("spokesmodel," as it turned out — a blonde babe who'd been hired two weeks prior), I stated flatly that the stuff was completely bogus and a total rip-off.

Liquid Trust immediately disappeared from the market. But a month later, it was back on. The last time I checked Google, there were seventy-six pages of ads and reviews, not just for Liquid Trust, but for a range of competitors that had taken the same shred of misinterpreted science and turned it into similar fairy-tale magic potions to exploit the gullible.

Which raises the question: Just how shameless do you have to be to sell a fake product that claims to create, of all things, trust?

There's nothing new, of course, about being shameless, or ruthless and cynical, when it comes to making a buck. Plenty of people in business seem to think that fakery and exploitation is the name of the game. Which is one reason that trade and commerce have always had something of an image problem. "Behind every fortune is a great crime" is one way of looking at it. "Never give a sucker an even break" is another.

Contrary to those sentiments, I'm going to show in this chapter that, on balance and despite its detractors, the marketplace actually makes people *more* moral, not less. Trade not only supports

159

oxytocin's virtuous cycle, it extends it beyond the small circumference of kinship or friendship. And then, with a twist that will come as a revelation to the "never give 'em a break" crowd, moral behavior actually *increases* the efficiency and profitability of trade. This adds another element to the virtuous cycle. A larger economic pie—also known as prosperity—reasonably well distributed, reduces stress and increases trust, which facilitates further release of oxytocin, which . . . you get the idea.

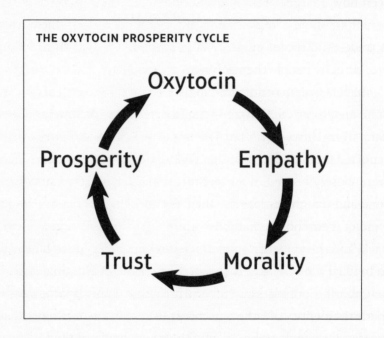

**THE OXYTOCIN PROSPERITY CYCLE**

Oxytocin

Empathy

Morality

Trust

Prosperity

These two threads can converge in making trade a positive moral force in the world, with the "action point" implication that the most sustainable markets—the ones we should be working to achieve and expand—are moral markets.

In keeping with our process so far, let's examine this proposition from the bottom up, looking at the biology that underlies market behavior.

In chapter 4 we talked about kin selection and how it was a driving force behind altruism in most social species. Social animals will look out for one another, even going so far as to sacrifice themselves for the good of the group. Doing that makes it more likely that the group will survive, and the genetic instructions to behave this way persist because the group's survival allows altruists' genes, including the "sacrifice-yourself" gene, to be carried along in their children, or even nieces or nephews. But we've also seen how a larger brain allowed our own species to discover the benefits of more complex forms of social cooperation. One of these is trade. And recent evidence suggests that the first thing traded was actually people themselves.

In 2011 a team of anthropologists led by Kim R. Hill of Arizona State and Robert S. Walker of the University of Missouri analyzed data from thirty-two contemporary hunter-gatherer tribes and reported that fewer than 10 percent of the members in each band were closely related. It turns out that this diversity results from sons and daughters leaving their family's band to join with the band of their chosen mate. Assuming that this tradition goes way back, and the evidence suggests that it does, we can see how each individual's blood relatives would become distributed among neighboring populations. Meanwhile, pair bonding would have made the identity of fathers more explicit, which would have made it easier for people to know who their widely distributed relatives were. All of which would have given members of neighboring bands a genetic incentive to cooperate with, rather than kill, each other.

But this same bias for out-breeding also meant that kin selection would have become a less potent force in promoting good behavior within each band itself, because not everyone in the band would be related by blood—there would also be all the new in-laws.

Which would have put much more of a premium on reciprocity—the exchange of favors—as well as on the need to preserve one's reputation for reciprocity, as the incentive for treating each other well.

## How Trade Enables Generosity

But the exchange of people among groups and the familiarity and trust it inspired also opened up greater opportunities for exchanging other things. Perhaps the tribe over there had a better technique for making arrowheads, while the tribe over here had a better technique for fashioning water gourds. One way to benefit from the exposure to diversity is to copy what the other guy is doing. But peaceful cooperation also meant being able to trade gourds for arrowheads, which meant that each group had the option of focusing and specializing. Which meant that not every band needed equal access to every resource in the environment, which meant that Tribe A didn't have to take from Tribe B in order to better themselves. Or as the nineteenth-century economist Frédéric Bastiat put it, "When goods cross frontiers, armies do not."

When trade came along, prosperity ceased to be a zero-sum game. In fact, more often than not, trade means that I'm often better off to the extent that *you* are better off. As a successful trading partner, you'll be feeding ideas back and forth with me—social learning—while also rewarding my efforts and sustaining my prosperity by paying me for what I produce.

In the past couple of decades, scientists have been able to explore the transition from primitive self-sufficiency to more market-oriented ways of making a living, using the same economic games we use in our OT studies to quantify moral behaviors.

Tools like the Ultimatum Game produce remarkably consistent

results anywhere in the world—provided that the test subjects are university students. In the Ultimatum Game, the most common offer worldwide is 50 percent of the pie, and proposed splits of less than 30 percent are almost always rejected.

But when scientists adapted the game for use with a tribe called the Machiguenga, the 50 percent and 30 percent norms that held up everywhere else disappeared. Among these slash-and-burn horticulturalists living in the southeastern Peruvian Amazon, offers averaged 26 percent of the pie, and less than 5 percent of offers were rejected. It seems that these isolated and self-sufficient people had a very different sense of what it means to share, and no sense of what it means to negotiate a win-win solution.

This anomaly prompted the MacArthur Foundation's Research Network on the Nature and Origin of Preferences to launch a hugely complicated venture led by Joseph Henrich, now at the University of British Columbia, Herb Gintis at the University of Massachusetts–Amherst, and Rob Boyd at UCLA. They selected fifteen cultures to study—small-scale pastoral, agrarian, or nomadic—ranging from hunter-gathers in the jungles of South America, to forager-horticulturalists in Papua New Guinea, like the group I visited, to herders in the high deserts of Mongolia, to whale hunters in eastern Indonesia. Some, like the Machiguenga, had no concept of trade—they killed or harvested everything they ate, and made everything they used. At the other extreme, some of the tribal peoples elsewhere, even though they still lived in the bush, occasionally took part-time jobs for wages. In the middle, other groups hunted game and harvested most of what they ate, but they also sold their agricultural products and occasionally bought food commodities or industrial goods.

It turned out that the Shona of Zimbabwe, for instance—who grow and sell commodity crops like maize, who produce pottery and handwoven baskets for sale, and who take commissions as blacksmiths or carvers—made much higher offers in the Ultimatum Game than the Hadza of Tanzania, who subsist almost entirely on hunting and foraging, just as their ancestors did ten thousand years ago. This distinction held true for each of the fifteen groups studied.

After a hugely rigorous process of data gathering and analysis, the scientists found a direct correlation between generous, pro-social behavior and the degree to which any given culture had been exposed to the marketplace. This exposure is called market integration, which is measured as the percentage of household calories purchased relative to the calories obtained directly from nature. Each 20 percent increase in market integration was associated with a 2 to 3 percentage point increase in offers in the Ultimatum Game.

But the researchers weren't easily convinced. They also analyzed one hundred other demographic, social, and economic factors that might have influenced this behavior. What they found was that only two factors mattered—market integration and membership in a world religion, either Christianity or Islam.

Stripped to its essentials, then, market exchange is a bit like coming together to worship a higher power, at least in the sense that it propels a positive feedback loop. A free and well-functioning market, after all, is about reciprocity, which means serving the needs of others so that they will reward you in kind. Repeated exchange, rather than a fly-by-night market based on "take the money and run," requires fulfilling the other person's trust in you, which means delivering what you promised, and at a price that allows both parties to benefit.

## Greed Is Good?

All this good news about the moral effects of the market begs the question: If trade is such a benign form of social cooperation, how did we move from the state of nature, in which market exchange cultivated virtue, to Liquid Trust, the collapse of Enron and AIG, the subprime real estate bubble, Bernie Madoff's greatest-ever Ponzi scheme, and Raj Rajaratnam's greatest-ever feat of insider trading?

Twenty-four hundred years before any of these calamities, Aristotle had already concluded that trade was destructive to virtue because it made us focus on money rather than on wisdom or on other people. Apparently the great philosopher was willing to overlook the fact that the marketplace was also the social hub of cities, the place where humans exchanged not only goods but ideas. Even in Athens during the Golden Age, the place for assembly, for political speeches, and for philosophical debate was the agora, the marketplace, the same place you went to buy a chicken to bring home for dinner.

But Aristotle was not alone in his suspicions. In the Confucian world of China, *sheng,* merchants, were only one step above social parasites because they didn't create anything tangible. The medieval church forbade lending money at interest (the law of Islam still does), and it strictly enforced the notion of the fair price, as opposed to the current notion of whatever the market will bear.

In the nineteenth century, Marxists became the harshest of market critics, going so far as to declare that all property was theft and that private entrepreneurs were enemies of the people. But Marxists have always been stuck in a zero-sum view of the world, as if it's only about distribution (to each, from each) rather than about expanding the pie for the good of all. (Which, as we'll see, does put the onus on capitalists to make sure that this benign force—the market—actually does what we say it does, which is to benefit everyone, rather than just a few sharp dealers at the top.)

In the 1960s, hippies turned away from buying and selling (at least until they discovered head shops) and tried to live on love, sharing everything. That spirit lives on in the Burning Man Festival held each year in the Nevada desert, an art and love fest where nothing can be bought or sold (but where just about anything and everything can be "gifted" to someone). When I attended Burning Man, the only formal markets were for coffee and ice (but you could buy all sorts of things underground).

"Markets are evil" is also one thread in the anti-globalization, No Logo movement that stages protests at big economic summits all over the world. In 2009 even the pope got in the act, issuing a pre-summit encyclical calling for the establishment of a "world political authority" to regulate the economy to make sure that it served the good of all and not just the fat cats. (Apparently, terms like *globalization* and *outsourcing* were a tough go for the Vatican scribes who had to translate them into Latin.)

With more than twenty-four hundred years of steady opposition to the idea of buying and selling, there has to be some basis for the complaints that business can corrupt virtue. In my view, the problem arises when individuals lose sight of what a sustainable market looks like. Some businesspeople actually *embrace* the idea that commerce is evil because they think that being cold and ruthless gives them a certain macho cachet, and that being macho—cold and ruthless—makes them more effective.

In Oliver Stone's two *Wall Street* movies, Michael Douglas played Gordon Gekko, the "baddest bad dude" to ever manage a hedge fund or pull off a hostile takeover. Years after he first created the role, Douglas told reporters how tired he was of drunken fund managers following him out of restaurants shouting Gekko's signature line—"Greed is good!"—then adding something to the effect of, "You got that right, bro. You the *man*."

Somehow these guys had missed the point that Gekko was the villain of the script, that "Greed is good" was meant to be ironic (even Orwellian, like "War Is Peace"), and that the movie was written as a cautionary tale about the dangers of money-grubbing sleaze.

Certainly the negative stereotypes of the market are reinforced by corporate leaders who pursue profit at any price, unafraid to taint the baby food, pollute the groundwater, play accounting games, or lay off a few thousand workers to add a dollar to their company's stock value. I'm sure the makers of Liquid Trust could offer plenty of rationalizations for what they're doing as just good ol' "American way" capitalism, even if it's being done from Bangalore or Belarus.

But in truth, you don't have to be a sleazeball to argue that moral teachings and the requirements of staying on top in a market economy are two very different categories. Ask any MBA or economics major—it's self-interest that governs human affairs, right? And especially commercial transactions. Steven Levitt and Stephen Dubner said it right there in their Introduction to *Freakonomics:* Morality represents the way *we would like* the world to work; economics represents how *it actually does work.* Can't argue with that.

Well, actually you can. I argue with that proposition just about every day.

Morality is not wishful thinking—it's biology, specifically, as

167

we now know, the biology of oxytocin. This means the behaviors that align with pro-social behavior, commonly called moral behavior, aren't adapted from a Sunday school lesson but are time-tested survival strategies, shaped by the harshest realist of all, natural selection.

## Penguins and Prosperity

Which brings us back yet again to the godfather of the hardheaded, rational science of economics, Adam Smith. When you read his work as a whole, as opposed to a few selected paragraphs, you find him making the case that the pursuit of self-interest can indeed benefit all, but only so long as it takes into account the mutual sympathy that leavens the contrary forces that are almost always at work in us, namely, greed and aggression.

If you saw the documentary *March of the Penguins,* then you know how the dads in this unlucky species spend the entire winter standing in temperatures way below zero, in the howling Antarctic winds, with an egg tucked between feet and belly fat. (By this point in the breeding cycle, the mothers have taken off for the Antarctic seas—warmer, but hardly St. Bart's—to recover from their pregnancies by chowing down on baby squid.) Essential to the males' survival, and the survival of their unhatched offspring, is the way these guys huddle together for warmth. But also essential is the way they *rotate* the huddle, so that everybody takes a turn on butt-freezing periphery, everybody gets a turn in the cozy warm center, and everybody moves equally through every level in between. Each individual penguin wants to stay warm and to hatch his chick—that's the self-interest part. But to stay warm he needs the group, because without the aggregated heat of all those bodies he and his future offspring would freeze. To keep the group

alive, and thus each individual alive, everybody has to play fair—
to cooperate. In this case, everybody takes his turn in the warm
center, and everyone spends time on the outer edge getting his tail
feathers frozen stiff.

With the penguins, pro-social behavior and the ultimate self-
interest of any individual (survival and reproduction) are indistin-
guishable. Their pro-social behavior, which melds individual
interest with the greater good, creates the virtuous cycle, then
reinforces it in an endless loop. That's the model for economic
behavior Smith was talking about.

As for humans, the study of our biology shows that we dance,
we become inspired by the mystery of some higher power, and
we exchange goods. That's just what humans do. Every culture
throughout history has created markets, and when they were
opposed, as was the case at Burning Man, they emerged underground as black markets.

In ancient times cities were built around temples, and as late
as the eighteenth century travelers in Europe or North America
would have known they were approaching a city when they saw
church spires on the horizon. But shortly thereafter, the telltale
urban landmark became billowing smokestacks and redbrick factories. During the Gilded Age before the First World War, histo-
rian Henry Adams observed that the market had superseded
religion as the central organizing principle of all modern societies.
The religious energy that had once motivated the building of great
cathedrals, he said, had morphed into a drive to invent and acquire.

Today each city's signature is a cluster of corporate skyscrapers
housing offices in which commodities are not necessarily invented,
designed, or produced, but financed, bought, and sold. As for
serving as an emblem for the culture, when the 9/11 terrorists
wanted to strike at the heart of American society, they didn't attack

St. Patrick's Cathedral or the Mormon Tabernacle—they attacked the World Trade Center.

And yet, setting aside direct loss of life, our economy and our society suffered as much or more damage as the result of a very different crash and burn in 2008, when those who revered markets all too much rode a wave of greed-driven excess right into the ground. By acting as if their own personal greed was good, markets were perfectly efficient, and "let the buyer beware" exempted them from any moral responsibility, these guys gave the strongest possible reinforcement to those who consider all markets corrupt.

These hard-core, largely unregulated double-dealers could have saved us all a lot of grief if they'd simply internalized one of the more salient passages from *The Theory of Moral Sentiments:* "How selfish soever man may be supposed, there are evidently some principles in his nature, which interest him in the fortunes of others, and render *their happiness necessary to him,* though he derives nothing from it, except the pleasure of seeing it."

There is plenty to criticize about the way the contemporary markets are run, but one fundamental truth stands out: Since it was first transformed and turbocharged by the industrial revolution and the individualistic values of the Protestant Reformation, the market has proved itself to be an unequaled means for building prosperity. Some would say that industrial capitalism has led only to crass materialism, but the evidence actually shows that, on balance, prosperity, like religion, contributes significantly to human health and happiness.

For example, in the United States, from 1600 to 2002, the average price-adjusted income increased 6,900 percent, the average length of life more than doubled from thirty-five to seventy-eight, and infant mortality fell from one-third of all children to less than five deaths in one thousand births today. Meanwhile the homicide rate fell 92 percent. During the same period in France and Holland, two countries that experienced a comparable increase in prosperity, the homicide rate fell 88 percent.

I recently analyzed data from the World Values Survey on the percentage of people who say it's important to teach your children to be tolerant, how trustworthy they thought others were, and average income across countries. I wanted to see how tolerance and trust (an indicator of morality) related to a country's income. As the graph below shows, tolerance and trust increase almost in lockstep with average incomes. There are exceptions, but as more people move away from subsistence-level incomes, their greater sense of security provides the luxury of being trusting

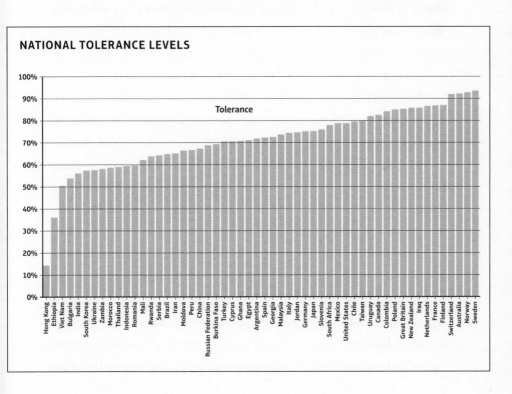

**NATIONAL TOLERANCE LEVELS**

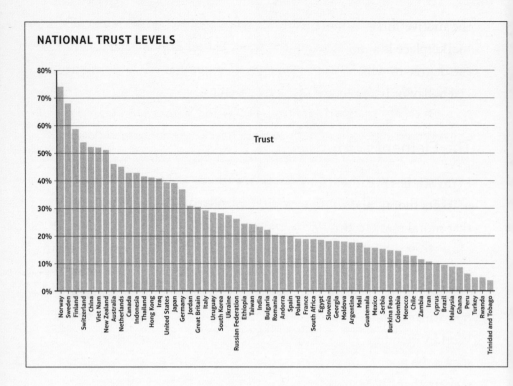

NATIONAL TRUST LEVELS

and tolerant. Research also has shown that tolerant countries are more innovative, producing the technological innovations needed to sustain prosperity.

While we can see the virtuous cycle impressively spinning along in the progression from poverty to prosperity, the question for developed societies is this: How do we preserve the self-reinforcing morality that is a starting point for successful markets and that produces such benefits? In other words, how do we sustain the prosperity that will produce the greatest happiness while protecting ourselves from a society that devolves into a winner-take-all banana republic, with $200,000 attack dogs as the new status symbol?

In the case of religion, we found that the balance tipped toward the positive to the degree that oxytocin trumped testosterone as the motive force. When it comes to the question of whether the marketplace is a moral positive or negative, the answer resides in the degree to which—not entirely surprising here—the commercial behavior is aligned with the release of oxytocin.

## The Elements of Economic Success

In my studies I've found four elements that are essential to keeping markets moral, and to deriving the maximum economic benefit that moral markets provide.

### 1. CONNECTION

A former graduate student of mine named Sherri Simms works for World Vision International, a nongovernmental organization that serves the poor in more than one hundred countries, and she had seen firsthand how violence and distrust prevent the accumulation of social and moral capital, locking countries into a never-ending poverty. She wanted to take the theoretical work I'd done comparing the effects of trust on prosperity and test it in the field, rather than through economic games in the lab.

Sherri knew that trade isn't just about goods and services but also the exchange of ideas, and social interaction in general, so for her dissertation we designed an experiment to see what happened when free Internet kiosks were introduced in six different villages on three continents.

Five of these villages were agrarian. The sixth, located little more than an hour's drive from Bangkok, Thailand, was a mix of agrarian and semi-industrial. With the introduction of the Internet, would people become glued to the screen and more isolated from their neighbors, or would connectivity contribute to the virtuous cycle as we saw when indigenous societies were exposed to markets?

Sherri was able to test villagers one month before and one month after they got Internet access that they used primarily to get weather reports and crop information. What we found was that, in each of the six settings, the rudimentary social exchange provided by Internet use increased trust as well as fifteen other measures of social capital. At the end of the experiment, in each of the six villages, there was more trust in others, more civic pride, more volunteering to help neighbors, and more overall satisfaction with life.

The evidence is pretty compelling that any connection that isn't abusive contributes to a positive feedback loop because connection builds trust. Given that the HOME system is constantly tuning itself to the environments in which we find ourselves, connection in one realm conditions us to cooperate in other realms, which ultimately can lead to a growth in prosperity, which then adds further to trust, which increases the willingness to behave generously and cooperatively.

In a highly unscientific study conducted for the magazine *Fast Company*, I did an experiment with a population of one, the business writer Adam Penenberg, to test the effects of social media as most Westerners experience it. While Adam was out in Claremont to do a story on my research, we took his blood before and after he spent fifteen minutes Tweeting. His oxytocin level increased by 13 percent, and his stress hormone ACTH decreased by 15 percent. It appears that even this most casual form of technologically mediated interaction—what psychologist Wendi Gardner calls "social snacking"—can have significantly positive effects.

In a replication of this experiment for the Korean Broadcasting Service, I tested peoples' blood before and after fifteen minutes of private and undirected social media usage and found that oxytocin increased for every person tested, and that the size of the change in oxytocin also correlated with the degree of connection. One young participant's oxytocin increased an astounding 150 percent. In my report to the KBS I speculated that he was online with either his girlfriend or his mother. They checked—he was posting to his girlfriend's Facebook page, and his brain processed the experience of connection as if she were in the room with him.

## 2. TRUST

When I was working with the chief psychiatrist of the San Diego Superior Court, Dr. Ansar Haroun, one of the inmates I tested, sitting in her orange jumpsuit and shackles, was a meth dealer. In the Ultimatum Game she came across as scrupulously fair, splitting the pie at exactly 50 percent. When I commented on this she said, "In my line of business, you cheat, you die."

The virtuous cycle isn't always enforced quite so ruthlessly, but very often the rule is, "You cheat . . . you're out of the game."

During the eleventh and twelfth centuries, the Maghribi traders of North Africa were far more successful than Genoese traders because they created bonds of trust that extended well beyond their kinship group. This made it feasible to engage local agents throughout the Mediterranean, and the key element was zero tolerance for cheating—one infraction and you were out, forever. So pursuing short-term gain through shenanigans was utterly foolish when it meant risking the lifelong benefits of the Maghribi network.

What the Maghribi knew from experience was that trust serves as an economic lubricant, lowering transaction costs by eliminating the need for elaborate systems of oversight and strict enforcement of cumbersome rules. Trust also provides such a compelling advantage in commerce that it then becomes a prompt to moral behavior elsewhere.

Orthodox Jews in the diamond markets of New York and Amsterdam operate according to the same principle—they don't even check to see what's inside the pouch when they hand over large sums of money. The same is true among the nine hundred high-end sushi wholesalers operating in Tokyo's massive outdoor Tsukiji market. Their reputation is on the line every day, with every batch of fish they sell, in terms of both quality and price. If they want to stay in business, they don't screw around or try to cut corners.

When people complain about having to fill out forms in triplicate and get twelve signatures whenever they're dealing with government, the root cause of the problem is lack of trust. In the nineteenth century, there was so much graft and corruption—it seemed nobody could be trusted when there wasn't an "owner" minding the store—that the crippling levels of oversight known as bureaucracy were instituted as a reform. Fortunately, after more than a century of bureaucratic bungling, people have come to realize that such top-down controls don't work, and believe it or not local, state, and even the federal governments have actually made great strides in modernizing with more bottom-up, market-based approaches.

The crippling effects of lack of trust become especially disturbing when we consider that the United States has been on a downward slope of trust since the 1960s, when 58 percent of Americans said that they trust others. Today that number is 34 percent.

The New York consulting firm Edelman publishes a trust ba-

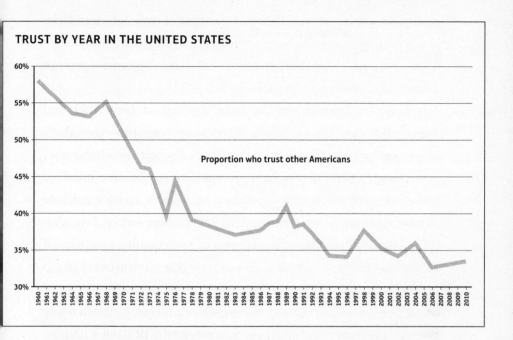

**TRUST BY YEAR IN THE UNITED STATES**

Proportion who trust other Americans

rometer that hit the lowest level yet reported in 2009. Sixty percent of the employees they surveyed said that they needed to hear information from a manager three to five times before they'd believe it.

Meanwhile, among the populace at large, a 2010 Associated Press poll shows that 50 percent of Americans have "little or no trust" in either corporations or Congress. The only institutions to inspire a "high degree of trust" were the military and small business. Our view of government and our view of business are alike in that trust appears to be inversely proportional to the distance from home—that is, we trust local government more than state, and state more than federal. In other words, we like to see a human face, and we like to know whom we're doing business with, and we prefer it if they have the same regional accent and root for the same sports teams we do.

## 3. A FOCUS ON SERVICE, AND QUALITY, NOT THE MONEY

At the business school at the University of Minnesota, psychologist Kathleen Vohs put two different groups of volunteers to work on computers. After a few minutes, subliminal messages appeared on the monitors in front of them. Ever so briefly, one group was subjected to images of fish sparkling under water. The other group was exposed to images of sparkling money. In psychology this is called priming, but in everyday parlance we know it as the power of suggestion. Even though participants had no conscious awareness of the images, this ever-so-subtle influence was enough to alter the behavior of the group cued to think about money. In tasks immediately thereafter, they were less helpful, less inclined to ask for help, and more likely to choose to work and play alone. When asked to set up chairs for an interview, the group primed to think about money also chose to put greater physical distance between themselves and other people.

The virtuous cycle induced by trade can be diminished anytime profit replaces people as the central concern. In a Mafia movie, when you hear the phrase, "It's not personal, it's business," you know that somebody's about to get whacked.

Which is why Frances Frei, Harvard Business School's expert on the service industry, reminds her students that the basic idea behind business is "to be of service." What a concept! Makes commerce sound kind of like a religious calling, doesn't it? Well, when you consider the good that it *can* bring, it's certainly not a bad aspiration. Serving others, as we've seen, causes the release of oxytocin and begins the virtuous cycle of moral behavior. Markets give us the chance to serve others every day.

All the same, global commerce and big finance, because they

so easily become abstract and impersonal, can easily be corrupted. That's one of the reasons that the lopsided growth in financial services has become such a problem for our economy. Mega-corporations like Walmart try to offset this impression by putting "greeters" at their doors.

Given the wrong conditions, even face-to-face transactions can become impersonal, and thus inhuman, with people seen as a commodity. Which, of course, opens up the market for competition on the basis of the human touch.

Before it was sold to Toronto-Dominion, Commerce Bank was the fastest-growing bank in the United States. It didn't give the best savings rates, and its range of services barely extended beyond providing checking accounts. But in a way, that was the whole point. It found its niche in simply being a retail bank, not a multipurpose, sophisticated financial powerhouse, and it won hands-down on the basis of having the best hours and the nicest people. Keeping offerings simple meant that the bank didn't need that many costly MBA financial wizards on hand to explain all the complexities— there were no complexities. As a matter of fact, the primary criterion for getting a job at Commerce was "does this person smile in a resting state."

Over the past couple of years I've gotten to know a group of businesspeople led by John Mackey, CEO of Whole Foods, who are trying to guard against the negatives in commerce by practicing what they call "conscious capitalism." These are hardly pinkos or head-in-the-clouds dreamers. Over the past ten years, conscious capitalism corporations have shown returns of 1,026 percent, compared with returns of 331 percent from business guru Jim Collins's "Good to Great" companies.

They begin by assuming that the first question to be answered is "What is your purpose?" What's so refreshing about this is that simply by asking the question, they challenge the assumption that the only reason you're in business is to make a buck—or, as it's expressed in current CEO lingo, "to maximize shareholder value."

After Hurricane Katrina devastated New Orleans, it took three weeks for Whole Foods to find all their employees. When they did, Mackey and his board decided to pay a full year's salary for all their New Orleans workers whether the stores were able to reopen or not.

This kind of concern for employees runs entirely counter to the master-of-the-universe approach that sees a CEO's task as beginning and ending with the quarterly return. The problem with the "maximize" approach is that it ignores the fact that every corporation has not just shareholders but also stakeholders, which include customers, workers, the community, and the society in which it operates. The idea that maximizing shareholder value is all there is to running a business leads to short-term thinking, which shortchanges the future and sometimes creates colossal blunders. Refreshingly, business leaders like Warren Buffett now argue against the idea of even providing quarterly reports on earnings. It's much better, they say, to extend the leader's gaze beyond the stock ticker and into the next five or ten or even twenty years. How is your business going to cope in a world where fossil fuels are no longer abundant? How are you going to adapt to increasing prosperity in Africa? You don't make these kinds of long-term adjustments without deploying your resources a little differently, moving some of your current effort away from exploiting the present moment and more toward exploring the future.

The conscious capitalism movement has much in common with the servant-leader model advocated by my late colleague Peter Drucker and business guru Ken Blanchard. The idea here is that the manager must see the people she leads not just as a means to an end, but also as an end in themselves. By engaging with employees on a human-to-human level, leaders tap into the HOME system in which human bonds, rather than fear or coercion, are the driving force behind effective collaboration and give-my-all productivity.

Business ethics is another arena in which oxytocin is surely the best possible guide, with the virtuous cycle as its own reward. When eBay was just a start-up and looking for new ways to expand, the company took in a huge investment from Bank of America. Within six months CEO Meg Whitman realized this new partnership was not viable, so she shut it down and immediately gave back all the money, even though that meant eBay would not be profitable for the year. At first Bank of America was shocked; it had already written off the money. In the ensuing years Bank of America has sent eBay so much business that eBay made far more in profits than the money it voluntarily returned.

But Meg Whitman tells another story even more in keeping with our theme. As eBay grew, it became increasingly problematic to translate ethical standards and standards of taste when expanding to other countries. The company had to draw the line somewhere as to what was or was not acceptable to sell on the site. The criterion it came up with was, "Would you be comfortable telling your mother you were doing this?" It translated perfectly across cultures.

## 4. EVERYONE BENEFITS

The most fundamental assumption of a consumer economy is that there will be plenty of consumers—people with money in their pockets and enough confidence in the future to spend it on all manner of goods. Just like penguins, consumers are predisposed to participate in social interactions, including trading goods. The term *free trade* is a misnomer if it's taken to mean free of rules, because all trade depends on rules. But fair trade is essential for non-coerced transactions. If markets produce a sense that you're not getting a fair shake, trade will evaporate. Trade is human. Trade is us.

Money definitely won't buy happiness, but Nobel laureate Daniel Kahneman reports that satisfaction with life continues to increase with income up to, and likely beyond, $160,000 a year. But you don't have to be making six figures to feel better. All other things being equal, life is just a little easier at $50,000 a year than it is at $15,000 a year.

And despite the fact that American society contains vast discrepancies in incomes and living standards, the power of our economic engine has created enough prosperity to mitigate many of the ways in which inequality can impair the virtuous cycle. Economic freedom (the ability to pursue one's economic goals without excessive government regulation) is itself strongly associated with happiness. Sociologist Jan Ott has reported that, despite economic hardships and frustrations in many circles, happiness in the United States has actually been rising. Not only that, but the inequality in happiness has been falling—meaning that the happiness gap between the richest rich and the poorest poor is less than it used to be, much of which has to do with non-economic gains that have been made by various groups. Social commentator Will Wilkinson has written, "If you want fewer materialists, make more material readily available to people, at which point they'll stop worrying about it so much and start worrying instead about things like happiness and the meaning of life."

That's why a market economy is very much like the penguin huddle I described earlier. It only works if there's enough of the heat provided by everybody else to go around, and if it's distributed evenly enough so that nobody is freezing his butt off.

Prosperity can be impaired either by excessive top-down controls; or by the absence of empathy that leads to winner-take-all conditions that in turn erode trust and the other, pro-social behaviors that follow from trust. When people worry about survival, it not only inhibits their oxytocin release, it impairs their consumer confidence, which has often been the first step in economic downturns.

To thrive long-term, any market—or business or society—requires fair, clear, and enforceable rules of exchange that sustain the virtuous cycle of trust, oxytocin release, and reciprocity. This doesn't simply make markets moral, it also makes them efficient—better at producing sustainable prosperity, the kind that doesn't burn itself up in golden toilet seats for the few but leads to an increasingly expanding economic pie.

But even the most minimally intrusive and appropriate fixes at the top won't be enough to maintain the kind of societal trust that can continue to drive sustainable prosperity. We also need a bottom-up approach that, through oxytocin release, taps into the virtuous cycle, and that removes the impediments to trust that corrupt and obstruct the cycle.

Just as we saw with the indigenous populations moving from subsistence survival toward an economy based on mutual trade, culture matters. It is now time to look at ways that each of us can work from the bottom up to shape our own culture so that it best reflects the wisdom of the Moral Molecule.

# CHAPTER 8

# A Long and Happy Life

*Mimes Creating Bottom-up Democracy*

Bogotá is a beautiful city that these days attracts throngs of tourists. But in the 1980s and early 1990s, you'd have to have been crazy to go there. Colombia's capital was caught up in the same kind of drug war we now see along the U.S.-Mexico border, with pitched battles in the streets among drug cartels, and between the cartels and the police. There are plenty of reasons why the police eventually won and the violence subsided, but a good part of the credit for bringing Bogotá back to life goes to Antanas Mockus, the philosophy professor turned mayor who, when it came to restoring civility, took a bottom-up approach to cracking down.

As part of his effort to rein in bad behavior, Mockus did something that at first blush sounds ridiculous—he put mimes on street corners. But it turns out that people fear public ridicule more than they fear a police citation, so there was method to the madness. When these performers poked fun at reckless drivers and jaywalking pedestrians, it actually made the scofflaws change their behavior. To break the grip of fear and distrust and to begin to rebuild Bogotá's damaged social fabric, Mockus also turned huge areas of the city into an enormous block party, curtailing auto traffic on Friday nights and Sundays. He launched a "Night for Women," urging the men to stay home and take care of the children while seven hundred thousand wives and mothers stepped out to celebrate, with female police officers on hand to maintain order. Addressing more mundane considerations, he took a shower on TV to demonstrate how to save water, and water usage dropped by 40 percent. Through similar gestures relying on humor and creativity rather than "thou shalts" or "thou shalt nots," he fostered empathy and built social—and moral—capital. He even got citizens to pay 10 percent extra in voluntary taxes. And voluntary disarmament days helped bring the murder rate down to one-quarter of its previous level.

Mockus's sometimes silly antics as mayor embodied the Confucian wisdom that "The great man is he who doesn't lose his child-heart." The broader lesson is that empathic human connection can succeed where top-down rules and the fear of punishment fail.

In the last chapter, we looked at ways that a human touch within the marketplace can elevate oxytocin and fuel the virtuous cycle. Here we're going to look at ways of achieving the same goal more broadly, across society as a whole. Once again, the key is the human engagement that causes oxytocin to surge, which increases empathy, which further increases human engagement.

I thought of Mockus and Bogotá the last time I was in New York City and saw the progress that's been made in turning those "mean streets" into places where human beings might want to spend time, maybe even contribute to the good feeling. Long stretches of Broadway have been turned into a pedestrian mall, with café tables and chairs welcoming people to sit and mingle and enjoy a festive atmosphere. The whole city has been spruced up and the crime rate is way down. Bryant Park and Madison Square Park, once not good for much but selling and shooting up drugs, now have outdoor dining, lights strung in the trees, and are packed with people well into the evening. The once bleak meatpacking district is now buzzing with tourists, and an innovative park has opened to rave reviews on the abandoned railway tracks running along the west side.

I also thought of Mockus's "child-heart" when I saw a front-page photograph of three tough-looking police officers in bullet-proof vests, sitting on the floor in a day care center in Río de Janeiro, bouncing diapered babies on their laps. This incongruous scene took place in one of Brazil's poorest and most stressed-out favelas, the infamous slum known as City of God, a district once so violent that the police simply retreated, leaving the people to the mercy of teenage gang members blasting away at one another with rocket launchers. Abandonment only increased the residents' distrust and resentment of the police, but so did the incredible brutality the authorities displayed whenever they did intervene. Then community policing arrived, with officers who not only visited day care centers to crawl around with babies, but also played soccer with the older kids and taught them to play the guitar and piano. Initially, these "police pacification units" had to be recruited right out of the academy to ensure that the officers would not yet have become corrupted by drug money. But once basic order was restored, the drug dealers lost control. Earthmoving trucks were able to come in and dredge the narrow, sewage-filled river, and garbage collection was instituted three times a week. The school dropout rate plummeted, with attendance at one high school showing a 90 percent rise.

Community policing goes back to the 1970s in efforts in cities such as Dallas and San Diego to foster cooperation and trust between citizens and police. This meant more foot and bike patrols and more minority officers in minority neighborhoods. By the time the concept reached New York, it also included zero tolerance for infractions against quality of life. The police became more proactive in enforcing laws against loud music, turnstile jumping, littering, and public drunkenness—the idea being that eliminating the minor infractions helps create a sense of community in which major crimes are less likely to occur. So the authorities were addressing both sides of the hormonal equation—cooperation and sanctioning—that we've been discussing all along.

Closer to home, I was enlisted to help the world's largest police force, the L.A. County Sheriff's Department, with an innovation they called trust-based policing. Sheriff Lee Baca and Lieutenant Mike Parker wanted to set up a system of shared leadership where line deputies were empowered to make decisions outside the typical, paramilitary command-and-control structure. This increased the degree to which deputies were responsible for their actions, but the union supported this because officer infractions would no longer result in the punishment of unpaid leave, but in required (paid) attendance at specific college classes to enhance job performance. Morale rose and infractions fell, and the sheriff enlisted community members in a program of listening and transparency that is now being copied by police forces around the world.

We know that relying on positive, personal contact rather than intimidation (emotionally remote officers in mirrored sunglasses patrolling in squad cars) works at the community level. But the question remains: How can we adapt the underlying concept of shifting toward a little more oxytocin and a little less testosterone to improve the workings of entire societies?

During the 2008 presidential primary season, we wanted to test the prospects for scaling up the human touch, so my team infused 130 volunteers with either a placebo or oxytocin, then tested their attitudes about trust. It came as no surprise, given the results of our earlier experiments, that the people on oxytocin expressed more faith in other individuals. But the oxytocin surge didn't just increase their trust in the people in the lab with them, it increased their trust in people in general. This hormonally induced change in outlook, in turn, led to greater trust in civic institutions, including government itself. It wasn't that oxytocin created more faith in any specific policies or political ideas. But those who received oxytocin showed increased trust in others who trusted the government. Which happens to be the foundation for democracy.

Then we decided to take it one step further. We wanted to see what effect, if any, oxytocin might have on people's political preferences. For this experiment we asked participants to describe themselves politically, either as an independent or as a member of one of the two major parties. We then infused them with oxytocin and presented them with a list of questions and propositions such as "rate your feelings toward Hillary Clinton" or "rate your feelings about Rudy Giuliani." Not too surprisingly, Democrats on oxytocin reported 30 percent greater warmth toward Hillary Clinton and 29 percent more warmth toward Congress. But Democrats infused with oxytocin also had more positive feelings about Republican candidates. Democrats on oxytocin showed 28 percent greater warmth toward Rudy Giuliani and 30 percent more warmth toward John McCain than Democrats infused with a placebo. Independents on oxytocin showed greater warmth toward both the Democratic and Republican parties, but not for any particular candidate. For Republicans, however—those who identified with the party for whom distrust of government is a core ideological position—the Moral Molecule had no effect whatsoever. Oxytocin did not increase their trust in Republican candidates, in Democratic candidates, in Congress, or in minority groups.

So what does this tell us? Well, it suggests that oxytocin can, indeed, serve to spark empathy and social connection at the individual level, which can then spread all the way up to the societal scale, but that oxytocin faces the same obstacles as a social force that it does as an interpersonal one. One of the factors that can short-circuit oxytocin is a deeply entrenched abstraction, whether that fixed idea is "rational self-interest," "those people are no damn good," or "government is the enemy."

Social observers, from Jane Jacobs (*The Death and Life of Great American Cities*) in the early 1960s to Robert Putnam (*Bowling Alone*) in our own era, have advocated building human capital by creating interconnected communities that operate on a human scale. Jacobs extolled the virtues of New York's Greenwich Village, where instead of having office parks here, subdivisions over there, and shopping malls a couple of exits down the freeway, people could live, work, play, worship, and shop—maybe even go to elementary school—all within a couple of city blocks. This jumbled concentration of activities allows people to know one another—and themselves—not just as workers, or as neighbors, or as parents, but as all these things, full-fledged human beings with all aspects of their lives coming together into an integrated whole.

In the 2010 elections in the United Kingdom, David Cameron campaigned on the idea of trying to create more of that "village" feeling throughout Britain—with village transparency, responsibility, and control—as a way of revitalizing the country. Elected prime minister, he pushed not only decentralization, local control, and charter schools—all familiar initiatives—but also the idea that individuals need to get out and rub shoulders with others and make things happen, not just in terms of financial self-reliance and entrepreneurship but in all aspects of citizenship, including charitable giving. In this he's harking back to the ancient Greeks, who had a word for those who did not take an active part in public life—*idiotes*. Guess what English word derives from it?

So far, the effort has not been warmly received by the British people, and lack of trust borne of the class system, rapid immigration, and economic stress—along with lack of a Mockus-style human touch to overcome those barriers—may have something to do with it. Even so, the theory behind the effort is grounded in solid science. As we've seen again and again, showing trust (and asking people to take on more personal responsibility is a sign of trust) builds trustworthiness, as well as empathy, generosity, and all the other forms of pro-social behavior we call moral. And equally important, research shows that in the post-industrial, globalized information economy, prosperity is based on the ability to navigate widely diverse social environments. To develop those skills, people need exposure to social networks where, even at an early age, they are called upon to be aware of what creates their well-being, to take responsibility for it, and to contribute directly to it.

Will Cameron's program succeed in time? Too soon to tell. Could anything like it ever work in the United States? Well, the U.S. is certainly a much larger country, with huge regional and ideological differences, which adds to the degree of difficulty. But here's what we do know: A nation's prosperity is directly correlated with trust, and trust is correlated with exposure to and engagement with others. The early work I did in economics, which led to my work with oxytocin, identified impediments to creating high-trust societies. Trust, and therefore prosperity, declines anytime vast discrepancies in income create barriers between people. The same is true for ethnic, religious, or linguistic differences, when they're allowed to stand as obstacles. Poverty, too, is a potent restraint on trust, with the stress of subsistence consumption inhibiting oxytocin's actions. In a recent study of sixty-eight hundred people in thirty-three countries, societies that come under threat also become less tolerant. So even at the societal level, when we really need to pull together, stress inhibits oxytocin release and gets in the way.

These societal effects dovetail with the obstacles to oxytocin release in individuals we've already discussed: genes, trauma, an over-reliance on reasoning to the exclusion of positive emotions, and perhaps the biggest culprit of all—testosterone and its behavioral repertoire of anger, hostility, and punishment.

Neuroscience gives us the fundamental ingredients we need to create a more oxytocin-rich, trusting, and prosperous society, but the policies we adopt to get there need to be developed within the political process. So what I want to do here is offer some thoughts on where we ought to steer, not dictate how we ought to row.

My research uncovered four important aids to this navigation.

## 1. ENHANCED COMMUNICATION

In order to develop, and then display, the trust and empathy that keep the virtuous cycle spinning toward trust and prosperity, we have to interact widely, and not just with people who look like us and think like us. My research has shown that one route to this destination is freedom of association and an unfettered media.

In this respect, those who want to foster civic engagement within the United Kingdom have an advantage in that so many of the people who run the government, the large corporations, and the large NGOs, as well as most of the people who comment on such activities in the media, can't help running into one another in the extended "village" (albeit a very large village) that is London. This means that adversaries are more likely to know each other, to actually see each other face-to-face more often, perhaps even running into each other out with their families on a Sunday afternoon.

The kind of face-to-face, informal interaction that tends to humanize people is harder to achieve in a nation of three hundred million people spread across a vast continent, and with many different cultural and political and economic centers. Since the nation's founding, Americans have tried to stitch this huge country together with the latest technology at hand: canals and riverboats, the pony express, the telegraph, the transcontinental railroad, the telephone, air travel, radio, television, the Interstate Highway System. Now the stitching is virtual, and increasingly global. When I was in the Papua New Guinea highlands, without electricity, without water supply or sanitation, I still had consistent mobile phone service provided by the Jamaican company Digicel. The tribal chief had a mobile phone, too.

When networked computing came along, broadcasting as a form of cultural glue was supplanted by multicasting, meaning that communication was no longer dominated by one source transmitting to everyone, but that messages could be pumped out by anyone to everyone.

An explosion in social media followed, and, as we've seen, even "social snacking" on Twitter or checking out a loved one's Facebook page can create the kind of oxytocin surge that increases trust.

New media is an incredibly potent force that has the potential to foster understanding throughout our society, and among all societies. But it needs to be wielded with care, and as in all things, the criterion for success is the extent to which what goes on actually widens, rather than narrows, the virtuous cycle. Is it oxytocin-driven or is it testosterone-driven? Does the communication foster human connection or does it foster anonymity and abstraction to the point that it cuts off empathy?

Providing the means for a billion voices to be heard—at least theoretically—as part of a global, electronic conversation 24/7 sounds like a great idea, but it doesn't necessarily lead to the Promised Land of a high-oxytocin, high-trust environment.

There's a problematic side to this, the first part of which might be called the Tower of Babel Problem, in which news and entertainment have become fractured into hundreds if not thousands of segments drowning the world in too much unfiltered, unreliable information. Then there is the Silo of Self-Absorption Problem, which allows individuals to tailor just about everything they see and hear, so that their entire experience online, as well as on radio and television, excludes anything that will truly expand their perspective, or challenge their preferences and prejudices. If you follow the commentary of Keith Olbermann or Bill O'Reilly exclusively because you like what you hear, you're reinforcing what you believe, but you're hardly getting the whole story. You can interact with hundreds of people all over the world every day in a jihadist chat group, or in a Christian chat group, or in a Kennedy assassination chat group, without ever encountering a thought that would connect you with anyone outside those groups.

During the Great Depression we had only newsreels and the radio, but there may have been a stronger sense of sharing a common reality in the United States then, when everyone—farm families in Alabama, recent immigrants in the Bronx, movie moguls in Malibu—sat down to listen to the president's "fireside chat." Similarly, I remember being struck by the social glue I saw in Brazil when I spent several months traveling there during graduate school. From the smallest towns in the Amazon to the financial capital São Paulo, everyone watched the same *futbol* game, news, and telenovela—then talked about it the next day. This provided a shared experience in a country larger than the continental United States.

Today customized media allows individuals to a large extent to create their own reality, which doesn't necessarily overlap with the larger reality that includes all their fellow citizens. It doesn't even extend throughout the same household, with different family members plugged into different media in different rooms. And, of course, every day we see the emblem of our age, which is three teenagers hanging out together, each exchanging text messages with someone not present.

In 2010 the Kaiser Family Foundation reported that Americans between the ages of eight and eighteen spend on average seven and a half hours a day using some sort of electronic device. The same year the Pew Research Center found that half of American teenagers sent fifty or more text messages a day, and that one-third sent more than a hundred a day. More than half said that they text their friends once a day, but only a third said they talk to their friends face-to-face on a daily basis.

Adolescence has always been a period of intense social activity, but developmentally it's also a period in which the wiring of the human brain is still a work in progress. We've already discussed how the HOME system is "tuned" by early interactions, and how friendships help kids build trust in people outside their families, laying the groundwork for healthy adult relationships.

Facebook, Google+, blogging, Tweeting, and texting make it

possible for less outgoing kids to mingle and develop certain social skills, which is wonderful. And many experts argue that smart phones and laptops, by enabling parents to spend more time at home, may result in more quality time between parent and child.

On the other hand, electronic communication is what psychologists call single-stranded interaction, meaning that it lacks the nuanced give-and-take that comes from social cues such as facial expressions and body language. Some neuroscientists worry that young "digital natives" are already having a harder time reading those social cues. (Even during the age of television, I remember schoolteachers saying that they now had to address each child individually, as in "Jenny, pull out your spelling book. Johnny, pull out your spelling book," if they wanted to get their attention. If they addressed the class as a whole, they encountered blank stares, as if the teacher in the front of the room were simply background noise, like a show the parents were watching while the kids were in the room.)

No one has proven it, but there's also the worry that lack of immediate feedback, as well as anonymity, may in some cases diminish empathy, which might account for the kind of cyberbullying that's already become a serious problem in teenage culture online. In my own home, the no-electronics rule applies even if we're just taking a drive. Just getting my family to talk to one another more works for us.

Simply put, while technology creates new opportunities for connection, it can sometimes provide new opportunities for neglect.

In her book *Alone Together,* Sherry Turkle, director of the Massachusetts Institute of Technology Initiative on Technology and Self, explores the effect on children of their parents' devotion to handheld electronic gadgets. She interviewed hundreds of kids who were very consistent in describing their feelings of hurt when their mom or dad paid more attention to their electronic devices than to them. They even cited the same three settings in which the gadgets were particularly intrusive and hurtful: at meals, during pickup after school or an extracurricular activity, and during sporting events. Turkle goes so far as to describe the parental plea, "Oh, just one more quick one honey," as being like the alcoholic pleading for just one more drink.

Will constant exposure to this kind of parental distraction affect the development of oxytocin receptors in today's children? Time will tell. But once again, we need to remember that the quality of quality time can be best measured by the amount of oxytocin being released. A child—or an adult—knows when you're with him and when you're physically present but distracted. Bottom line: New media can bring us together in new and enriching conversations, or they can also send us off into our own private worlds, muttering to ourselves like crazy people on street corners, or ranting like true believers on talk radio. To get it right, we need to be sure it's genuine connection we're actually pursuing.

## 2. POSITIVE EXPOSURE TO DIVERSITY

Positive exposure to those outside our own family or cultural or geographic "tribe" is another element we need in order to achieve a more oxytocin-driven, pro-social, and prosperous society. This is all the more urgent because there are solid, evolutionary reasons why our species developed the tendency to be wary of those with physical appearances or behavioral patterns different from our own. After all, for millions of years an individual's social world was limited almost entirely to her village and tribe, and outsiders were, for good reason, considered a threat until proven otherwise.

Harvard psychologist Mahzarin Banaji has shown how deeply embedded these preferences are. In her studies, white babies prefer white faces to black faces from the earliest moment that it's possible to measure. But she's also shown that white babies who've been exposed to black faces early on lose the bias. In fact, white babies who've been exposed to black faces will, if they're accustomed to hearing English, demonstrate more comfort with a *black* English speaker than with a white person speaking, say, Norwegian. In other words, there's a self-protective basis for being wary of difference, but the suspicion is malleable, and it fades with exposure.

Today new waves of immigration add new challenges to these ancient biases, leaving established populations all over the world feeling overwhelmed by newcomers and the speed of cultural change.

In Europe, being accepted as part of the nation has a lot more to do with culture and ethnicity than it does in the United States. In France or Germany there is no tradition of the melting pot, no Ellis Island or Statue of Liberty in the harbor in Marseilles or Hamburg to welcome the huddled masses. And yet the huddled masses are flooding in from former colonies and from other economically and politically distressed regions. The French struggle to deal with their large Arab population, while the Germans struggle to assimilate the Turks who came as "guest workers" and stayed on. Meanwhile, powerful anti-immigrant movements have gained traction everywhere, even in liberal Scandinavia.

My research shows that, in the short run, immigration reduces trust, but that this negativity is mitigated as the newcomers assimilate. Trouble is, when immigrants meet with too much hostility, they remain apart, as has happened in Germany, where third-generation "German" Turks are more adamantly "Turkish" than their parents. Positions harden, and hostility breeds hostility, as each side feels threatened by the other.

Then again, my studies also show that diversity increases the variety of ideas and ways of doing things that can stimulate innovation. Moreover, acceptance breeds acceptance. Seventy years ago the United States was locking up ethnic Japanese in internment camps, even those who were citizens, even those whose sons were serving in the U.S. Army in World War II! Now, especially on the West Coast, being of Japanese descent is about as exotic as being Irish or Polish in Chicago, and Japanese Americans have proven themselves valuable citizens and contributors to the economy.

Again, it's the size and rate of demographic change that largely determine how people respond to difference. To cross the divide, we need the child-heart of bottom-up, interpersonal connection, unimpeded by negative ideas about racial or ethnic difference. And there's hope, even in what seems like the toughest cases, because attitudes are often more complex than they first appear. Arizona has gained a lot of attention for its controversial get-tough policy on illegal immigrants. Yet when it comes to international refugees, only three states welcomed more immigrants on a per-capita basis over the past six years than Arizona. Per capita, Arizona took in nearly twice as many people from Somalia, Myanmar, Iraq, Bosnia, and Sudan as did California, and more than twice as many as New York, New Jersey, and Connecticut. It's the scope of each demographic trend that tells the tale: Arizona's intake of refugees in 2009 was forty-seven hundred people. Its population of illegal immigrants is thought to be approaching four hundred thousand.

Given the size of these numbers, it's not entirely surprising that Anglos in Arizona fear that the stretch of desert where they live is being re-annexed by Mexico, and that they will become the outsiders. But the other issue is rule-following, because Arizonans take a dim view of anyone they see as "jumping the line." So here again, the oxytocin-driven impulse for empathy—help the refugees—is offset by the testosterone-driven impulse to sanction those who circumvent the rules and regulations.

To arrive at the proper course, Arizonans probably need to follow the lead of Antanas Mockus toward less fear and more fiestas.

Even when we're talking about people whose families have been U.S. citizens for generations, perhaps centuries, it appears that we are in a period of intense regional, cultural, and political divisiveness in which some bottom-up oxytocin could help. It's a problem when well-traveled Americans from the coasts, people who know their way around Tuscany and Provence and maybe Thailand, have never set foot in the giant landmass that lies between the Sierra Nevada and the Hudson River. Or when upscale business travelers refer to the "fly-over cities," expressing a feeling of disdain that is more than reciprocated by those feeling looked down upon by the highfliers passing overhead. It's not surprising that resentment of "elites" has become a hot button in certain political circles. The "just folks" out in the small towns and the farming country have returned the insult based on their own sense of injury by launching a rhetorical, and divisive, battle over who is, and who is not, a real American.

By the same token, the kinds of kids who do sailing camp or Outward Bound over the summers, then attend elite universities, often have little or any exposure to anyone who spent every summer bagging groceries at the A&P, then served in the military in order to pay for college. The small-town families who provide the bulk of military personnel often have little if any exposure to the cosmopolitan culture and values of the big cities.

For all these reasons, it occurs to me that a domestic exchange-student program might be in order, allowing prep school kids and small-town and rural kids to get to know one another and experience one another's lives. The need for this becomes all the more apparent when you think about just how hard it would be to pull off. The son or daughter of a lawyer from Paris would no doubt have far less trouble fitting in on the Upper East Side of Manhattan than would a kid from a farm family living outside Manhattan, Kansas. The language barrier is nothing, but the domestic cultural barrier is huge. A model for this sort of exchange is the Seeds of Peace summer camp in Maine, which brings together Israeli and Palestinian teens. In just a few weeks, they manage to build ties among these initially wary youths that contribute to positive change and can last a lifetime.

William Greider summed up the demographic state of play a few years ago with his book title, *One World, Ready or Not*. But there's an even older phrase that carries the spirit of oxytocin, which is the spirit we need: "We're all in this together."

## 3. PROCEDURAL FAIRNESS

Since 1789 the primary glue that's held American society together has been the U.S. Constitution, a few simple rules that can be adapted to changing circumstances but that, most importantly, ensure procedural fairness, institutional integrity, and transparency. It's only by our common agreement to uphold those values that we've been able to create and maintain the kind of trust that has allowed such a heterogeneous nation to thrive. The Constitution encourages trust by providing equality under the law, an impartial judiciary, freedom of the press and assembly, and the light to moderate economic regulation that allows the overall economic pie to expand. According to tradition, all this comes together to provide the sine qua non of a successful market-based society—a tradition of upward mobility based on merit.

But simply making speeches congratulating ourselves about the American Dream won't keep the oxytocin flowing, nor the virtuous cycle spinning virtuously for the benefit of all. During the past forty years the United States has become two separate societies based on income. This is the formula for a banana republic, with gate-guarded communities and private security forces, more than it is the formula for a society in which trust enhances prosperity.

The income gap in America is epitomized by the differential between average CEO pay and the pay of the average worker. Forty years ago it was eleven to one. Now it's four hundred to one. According to the Bureau of Labor Statistics, in 2010 median CEO salaries jumped 27 percent while overall worker pay increased by just 2.1 percent. America is now a place where the wealthiest 1 percent of the population control 38 percent of the country's privately held assets.

There's always been a trade-off between the need to provide for opportunity and growth and the amount of inequality we can tolerate. My cross-country research shows that providing short-term income support (a safety net) for the very poorest in society raises trust and benefits everyone. It also reduces crime. But too much support could move us back into shackling generation after generation into welfare dependency. The empathic approach is, I believe, not only to provide help for those acutely and adversely affected by the economy but also to give them a ladder out of poverty, which doesn't just mean a minimum-wage job putting french fries in a bag. It may require training in appropriate hygiene, as well as in concepts like showing up on time and showing up on Monday even though you got paid on Friday. For some, it might mean psychological counseling or appropriate medications.

This stuff doesn't come cheap, and there are probably as many different ideas for achieving the right balance as there are economists and political pundits, but one approach to keeping trust alive by keeping opportunity alive stands out as unassailable, and that is to focus on the fourth coordinate on our map—educational achievement.

## 4. EDUCATION

My research shows that improving the quality of education is a cost-free way to raise prosperity. It's cost-free because it reinforces so many of the other things we need to keep the virtuous cycle rolling that, ultimately, the increase in economic benefits far outstrips the cost of the investment. Education brings more people into the comfort zone of higher income, which increases trust, then causes people to demand better government, which further increases trust, which further reduces inequality, which increases the pool of those who will get a good education.

The promise to make public schools do a better job is one of the hardy perennials of American politics. But the data show that the most significant determinant of whether or not kids achieve their educational potential is bottom up—namely, whether or not they have stability and love at home. It's also true that parents who are motivated to truly invest in their children demand better schools.

Recently educational reforms have taken the virtuous cycle to heart in trying to inculcate positive emotions such as empathy. Educators are even experimenting with software that helps reduce stress and facilitates interpersonal connection. But when it comes to increasing empathy, there's a tradition that goes back a couple of thousand years that has been pretty successful in humanizing people. It's called high-quality exposure to the humanities—literature, foreign languages, philosophy, history, music, and art—all the stuff (now sometimes derided as "useless") that was once the common currency of any educated person. While studying the humanities is often knocked for being impractical, we need to remember that the oxytocin system is tuned and enhanced anytime we go inside another person's head by reading a good novel or listening to a sonata, or when we develop an understanding of another culture, or another historical epoch. So while we need to provide people with technical skills that will help them find employment, we can't afford to neglect the even more basic skills—reading, writing, thinking, feeling—that allow them to become fully realized human beings who care about the world they live in and the people who share it with them.

In 2011 a report came out arguing that keeping opportunity, and thereby trust, alive in this country isn't just a virtuous idea for well-meaning people—it's become a strategic necessity. And this analysis didn't come from some softhearted, academic think tank. It came from a colonel in the marines and a captain in the navy, both officers on the staff of Admiral Mike Mullen, the chairman of the Joint Chiefs of Staff. The United States, they maintained, can no longer afford to engage the world primarily through military force, and the only way to maintain our dominant position in the world, they said, is through the strength of our educational system and our social policies. According to these military strategists, our first priority should be "intellectual capital and a sustainable infrastructure of education, health and social services to provide for the continuing growth of America's youth." Meanwhile, the Department of Defense has begun to put its money where its analysis is, funding research in the neuroscience of social and moral capital, and I'm proud that my lab is one of those it's chosen to support.

## The Happiest Place on Earth

One country in particular is way ahead of us in implementing the kind of strategic vision that was recommended to the Joint Chiefs of Staff, and it's not China or India—the economic competitors we usually worry about. It's Costa Rica. We don't usually think of this Central American nation as having anything to teach us, but when you consider what they've accomplished, the results are pretty impressive. Sixty years ago they made the decision to abolish their army and focus their resources on education. Since then, they've enjoyed a more stable society than any of their neighbors, seen their economy thrive, and seen life expectancy advance until it is on par with that in the United States.

It's also true that, based on Gallup surveys and a database compiled by Dutch sociologists, Costa Rica—not Disneyland—is the "happiest place on earth." Compared to 148 other nations surveyed regarding their sense of well-being, Costa Rica came in number one. On a 10-point scale, Costa Ricans rank themselves on average 8.5. Denmark came in second at 8.3. The United States ranked twentieth at 7.4. Tanzania was at the bottom at 2.6.

Years ago, when I first looked at cross-country measures of trust, I investigated eighty-five variables that I thought might be associated with oxytocin release, testosterone, and stress at the societal level. The strongest correlation I found among all of these variables was the association between happiness and trust. This tight correlation continued to hold regardless of a country's income level. Rich or poor, living in a trusting society simply makes people happier.

Interesting, but should we as a great nation be concerned about happiness? Oddly enough, the Founding Fathers—hardly New Age fools—listed "pursuit of happiness" in the Declaration of Independence right up there along with "life" and "liberty" as one of the "inalienable" Rights of Man. And there's more to happiness than meets the eye.

Commencement speakers sometimes make the distinction between *happiness* and *satisfaction,* suggesting that the former is merely an upbeat mood, or the temporary fulfillment of a craving, while the other refers to pleasures that are longer-term, deeper, and more meaningful. You're "happy" when you find a parking space. You feel "satisfaction" when you've worked hard and saved and launched your children into successful adulthood.

As usual, the Greeks had a word that probably best nails what we're after. That word is *eudaimonia,* which means "to flourish," and it makes clear that the good stuff we're looking for—and that often goes by the name of happiness—is not just the transient or superficial satisfaction of an appetite, but a pervasive state of well-being, a condition that affects our entire physiology, including improvements in our immune system that can lead to a longer, healthier life, as well as greater all-around prosperity. *Eudaimonia* is "the good life" as defined not by Donald Trump but by the philosophers who set the cornerstones of Western culture.

Back in the "Gordon Gekko" 1980s there was a popular bumper sticker that said HE WHO DIES WITH THE MOST TOYS WINS. I doubt that anybody actually believed that, but, unfortunately, plenty of people still live as if they do. And this despite the fact that, as survey data tell us, the daily activities most often associated with happiness are pretty simple. *Eudaimonia* isn't derived from owning a $6,000 shower curtain or drinking a $400 bottle of wine. The things people rate most highly are having a good romantic relationship and many friendships, having a job you like, enjoying the community you live in, and having a level of income good enough to reduce the stress of just getting by.

Martin Seligman, pioneering researcher into human happiness, says that *eudaimonia* consists of five things: positive emotions, engagement, relationships, meaning, and accomplishment. Which sort of harks back to Freud's even simpler formulation for what people need—love and work.

Another leading happiness researcher, Arthur C. Brooks, puts the most emphasis on what he calls "earned success," which, again, suggests a long-term proposition, but still doesn't necessarily have anything to do with making money and waving it in other people's faces. The success he's talking about could be building a company or qualifying as a thoracic surgeon. But it could also be growing really great tomatoes in your backyard, or learning to play the banjo.

Aristotle, another guy who was hardly an airhead, based his entire system of ethics on *eudaimonia,* saying that the reason to strive for virtue is that being virtuous makes us happy.

In 2010 I decided to put Aristotle's idea to the test by running yet another variation on the Trust Game. In this study of sixty college-aged women, the participants were all B-players, and, unbeknownst to them, each received the same amount—$24—transferred from an A-player who was in on the game. We set it up this way because, this time, what we were after wasn't to determine the size of response in relation to size of stimulus, but to provide a consistent stimulus. This way, we could find out how women who release lots of oxytocin might differ from those who release little or none.

Before we got going, we put all the players through a series of surveys and tests that would indicate how they felt about their lives. We were then able to correlate the results from the Trust Game— their oxytocin surge and their resultant generosity toward the stranger who trusted them—with their answers to the questions, their baseline indicators of well-being, or *eudaimonia*. We found that those who had the largest surge in oxytocin were not only the ones who returned the most money but also the ones who'd reported greater satisfaction with life, greater resilience to adverse events, and lower scores for depressive symptoms. Those who returned the most money—the generous ones, maybe even the virtuous ones— left the lab with the least coin in their own pockets, but they were by far the happiest. And these oxytocin-adepts were all about connection. They had higher-quality romantic relationships (which resulted in *more* sex with *fewer* partners), had more friends, had closer family relationships, and were more generous to strangers.

Oxytocin, then, is not only tied in with the brain mechanisms that make us pro-social and moral, it is also tied in with the mechanisms that make us happy by activating the elements in the HOME circuit, dopamine and serotonin. Fulfilling relationships make us happy, and as psychologists and epidemiologists have been demonstrating for many years, being happy makes us healthier. Oxytocin reduces cardiovascular stress and improves the immune system, a neat trick for a tiny and ancient molecule, causing us to live not only happier, but longer.

Loma Linda, California, where I happen to reside, is the only so-called blue zone in the United States—a place where people commonly live past one hundred. When we did a study with these "oldest old," having them watch the "Ben's Story" video and then taking their blood, their oxytocin levels went through the roof. These oldest—and healthiest—people in America are also some of the kindest people you could ever want to meet. Just as we found among the college-aged women, those from this group who released the most oxytocin after the video also reported greater satisfaction with their lives, were more grateful for what they had, showed more empathic concern for others, and had fewer depressive symptoms.

It turned out that most of them had spent their lives working in professions that help other people, such as teaching and nursing. Curiously, even in this quite religious Seventh-day Adventist community, those who released the most oxytocin were actually the least religious. It was as if their connection to other people was so strong that it satisfied the craving that often leads people to try to connect with God.

Bottom line, their good health and joy in living such a long and happy life should be all the endorsement you'd ever need for letting yourself be guided by the Moral Molecule.

While writing this book, I asked my friend and colleague Earl Quijada, M.D., if I could accompany him on his hospice care rounds. I wanted to see if living a fulfilled life would lead to a better death. What I saw in these encounters offers a chilling cautionary tale. Earl makes house calls and coordinates a team of nurses, social workers, and chaplains to provide end-of-life treatment that tries to take care of the whole person. Sadly, some people have remained woefully undeveloped in the parts that allow them to connect and to experience joy. And their last days are not pretty.

"Hank" was one of these, a seventy-two-year-old who had end-stage Parkinson's disease. With a Ph.D. and an M.D., he had worked for forty years as a physician, first as an internist. But Hank had such poor people skills that sitting at a laboratory bench with a microscope suited him much better than working directly with patients, so he became a pathologist. When I saw him, not long before he died, he held his hands clenched as if he were beginning the long slide back into the fetal position, and he weighed about eighty-five pounds. He'd never trusted others, especially other doctors, so he'd served as his own physician until he became bedridden. At that point nursing aides had been brought in to help, but his physical and emotional outbursts drove them away. When I saw him, he was paying a neighbor and her twenty-year-old son to keep him clean and to bring him what he needed to get by.

The interior of his home helped explain how he came to such a dismal end. He had never married and did not have any children. There were no pictures of anyone, anywhere. Not a single sign of a single human connection. He died a day after I saw him.

Another patient I saw, "José," had end-stage heart failure and told me with a wink that he kept on surprising Earl by living so long. He was bedridden and weak of body, but he was sharp of mind and had a wicked sense of humor. His wife had planted a rose garden outside his window to give him joy even though he could no longer garden, and he proudly showed it to me. His room was filled with pictures of his children and grandchildren. While we were at his house, his daughter stopped by to see him, and José told me his son visited every night. José was not religious, but he told me frankly that he was fully at peace with dying and had lived and loved well. His only regret? In the last several months he was too weak to go to the park to see his grandkids play.

## Me-Search

In the nineteenth and twentieth centuries, economics tried to achieve scientific rigor by cutting off recognition of the human element of motives, expectations, and psychological uncertainties. Fortunately, behavioral economics, and now neuroeconomics, has put us back on what I consider the right track, which is a path that combines both rigor and moral perspective.

Alfred Marshall, a major architect of hardheaded, quantitative economics, encouraged his colleagues to "increase the number of those [in] the world with cool heads but warm hearts, willing to give at least some of their best powers to grappling with the social suffering around them."

I'm very fortunate that I've found a way to study the human element in all its glory.

There's a saying that "all research is me-search," and it may have been that the empathy-deficient environment where I spent so many years studying economics made me become so engaged with the study of oxytocin, connection, and morality. Now I'm definitely making up for lost time.

I hug everybody. A few years ago, I began warning anyone who visits my lab that before they leave I'm going to give them a hug. Even though this scares some people—especially economists—I've found that this slightly eccentric announcement changes the depth of the conversation, making it more intimate, more engaging, and more valuable to us both. People start to open up. I suspect that by forecasting a hug, I'm also signaling how much I trust the person, so I'm inducing oxytocin release in their brains.

My penchant for hugging everyone led *Fast Company* magazine to anoint me "Dr. Love" after I hugged their writer Adam Penenberg. So let Dr. Love offer you a prescription: eight hugs a day. We've shown that if you give eight hugs a day you'll be happier, and the world will be a better place because you'll be causing others' brains to release oxytocin. They, in turn, will connect better to others, treat them more generously, causing oxytocin release . . . yes, the virtuous cycle begins with a hug. The other thing I do when anyone comes to see me is to ask how I can make their visit with me the most valuable and fulfilling. This is part of being fully present and available, which is another lesson I've learned from the Moral Molecule.

I try to pursue that same wisdom in my own daily life, and I think it's helped me become a better teacher, team leader, husband, and father. For sure it's made me a much happier person. Most of the changes I've made are little things—like getting a dog for my kids and spending a lot more time playing with them.

I can't swear that this shift has tuned up my oxytocin receptors, but I do know that, as a six-foot-four former jock, gearhead, and recovering math nerd, I now love nothing more than piling up on the couch with my wife and two daughters and getting all weepy at movies about little girls and talking rabbits—something I never would have imagined when I was playing football or working on cars in my teens.

The "thou shalt" religious devotion that my mother tried to pound into me faded away a long time ago, but ironically, something at the core has remained. Oxytocin—a reproductive hormone—makes us moral, so ultimately, you could say that we are moral because of our origins as sexual creatures. Which harks back to that very Christian-sounding idea that God is love, or maybe that love is God. But as we saw, *eros*—sex—is only one kind of love, and oxytocin covers all the bases. Oxytocin makes us feel the love for others that's known as *philia,* the familial love known as *storge,* as well as *agape*—the love of the divine that we seek through self-transcendence, which can be released during dance, meditation, and magic.

My mother's faith also held that "the kingdom of God is within you," which is, at its root, a very bottom-up kind of idea. God is love. God is within you. Oxytocin is love. Oxytocin is within you.

So actually, the ancient sages were spot-on. Empathic human connection, governed by oxytocin, is the key to trust, love, and prosperity. It is the goodness that we seek.

# Notes

Bibliographic references are organized below by chapter.

## INTRODUCTION

Geddes, L. (13 February 2010). With this test tube I thee wed. *New Scientist.*

## CHAPTER 1: The Trust Game

Zak, P. J. (June 2008). The neurobiology of trust. *Scientific American,* 88–95.

Zak, P. J. (2011). The physiology of moral sentiments. *Journal of Economic Behavior & Organization* 77, 53–65.

Zak, P. J., Kurzban, R., & Matzner, W. T. (2004). The neurobiology of trust. *Annals of the New York Academy of Sciences* 1032, 224–27.

Zak, P. J., Kurzban, R., & Matzner, W. T. (2005). Oxytocin is associated with human trustworthiness. *Hormones and Behavior* 48, 522–27.

Smith, V. L. (1998). The two faces of Adam Smith. *Southern Economic Journal* 65, 1–19.

Fisher, H. (1994). *Anatomy of Love: A Natural History of Mating, Marriage, and Why We Stray.* New York: Ballantine Books.

## CHAPTER 2: Lobsters in Love

Donaldson, Z. R., & Young, L. J. (2008). Oxytocin, vasopressin, and the neurogenetics of sociality. *Science* 322, 900–904.

Zahavi, A., & Zahavi, A. (June 3, 1999). *The Handicap Principle: A Missing Piece of Darwin's Puzzle.* New York: Oxford University Press.

Carter, C. S., & Getz, L. L. (1993). Monogamy and the prairie vole. *Scientific American* 268, 100–106.

Ross, H. E., et al. (2009). Variation in oxytocin receptor density in the nucleus accumbens has differential effects on affiliative behaviors in monogamous and polygamous voles. *Journal of Neuroscience* 29(5), 1312–18.

Pitkow, L. J., et al. (2001). Facilitation of affiliation and pair-bond formation by vasopressin receptor gene transfer into the ventral forebrain of a monogamous vole. *Journal of Neuroscience* 21(18), 7392–96.

Kosfeld, M., et al. (2005). Oxytocin increases trust in humans. *Nature* 435(2), 673–76.

Zak, P. J., Stanton, A. A., & Ahmadi, S. (2007). Oxytocin increases generosity in humans. *Public Library of Science ONE* 2(11), e1128. doi:10.1371/journal. pone.0001128.

Morhenn, V. B., et al. (2008). Monetary sacrifice among strangers is mediated by endogenous oxytocin release after physical contact. *Evolution and Human Behavior* 29, 375–83.

## CHAPTER 3: Feeling Oxytocin

Barraza, J. A., & Zak, P. J. (2009). Empathy toward strangers triggers oxytocin release and subsequent generosity. *Annals of the New York Academy of Sciences* 1167, 182–89.

Smith, A. (1759/1982). *The Theory of Moral Sentiments* (Vol. I of the Glasgow Edition of the Works and Correspondence of Adam Smith). Indianapolis: Liberty Fund.

Pellegrino, G. d., et al. (1992). Understanding motor events: a neurophysiological study. *Experimental Brain Research* 91(1), 176–80.

Umiltà, M. A., et al. (2001). I know what you are doing: a neurophysiological study. *Neuron* 31, 155–65.

Rizzolatti, G., & Craighero, L. (2004). The mirror-neuron system. *Annual Review of Neuroscience* 27, 169–92.

Taylor, C. E., & McGuire, M. T. (1988). Reciprocal altruism: fifteen years later. *Ethology and Sociobiology 9*, 67–72.

Preston, S. D., & de Waal, F. B. M. (2002). Empathy: its ultimate and proximate bases. *Behavioral and Brain Sciences 25*, 1–72.

Humphrey, N. K. (1992). *A History of the Mind: Evolution and the Birth of Consciousness.* New York: Simon and Schuster.

Hrdy, S. B. (1999). *Mother Nature: A History of Mothers, Infants, and Natural Selection.* New York: Pantheon.

Meltzoff, A. N., & Moore, M. K. (1977). Imitation of facial and manual gestures by human neonates. *Science* 198, 75–78; Myowa-Yamakoski, M., Tomonaga, M., Tanaka, M., & Matsuzawa, T. (2004). Imitation in neonatal chimpanzees. *Development Science 7*, 437–42.

Preston, S. D., & de Waal, F. B. M. (2002). Empathy: its ultimate and proximate bases. *Behavioral and Brain Sciences 25*, 1–72.

Hatfield, E., Cacioppo, J. T., & Rapson, R. L. (1994). *Emotional Contagion.* New York: Cambridge University Press.

Buccino, G., et al. (2001). Action observation activates premotor and parietal areas in a somatotopic manner: an fMRI study. *European Journal of Neuroscience 13*, 400–404.

France, M. L., & Broadbent, M. (1976). Group rapport: posture sharing as a nonverbal indicator. *Group and Organization Studies 1*, 328–33.

Byrne, D. (1971). *The Attraction Paradigm.* New York: Academic Press.

Maurer, R. E., & Tindall, J. H. (1983). Effect of postural congruence on client's perception of counselor empathy. *Journal of Counseling Psychology 30*, 158–63; Lakin, J. L., & Chartrand, T. L. (2003). Using nonconscious behavioral mimicry to create affiliation and rapport. *Psychological Science 14*, 334–39.

Bernieri, F. J. (1988). Coordinated movement and rapport in teacher-student interactions. *Journal of Nonverbal Behavior 12*(2), 120–38.

## CHAPTER 4: Bad Boys

De Waal, F. (2005). *Our Inner Ape.* New York: Riverhead Books.

Zak, P. J., et al. (2005). The neuroeconomics of distrust: sex differences in behavior and physiology. *American Economic Review Papers and Proceedings* 95(2), 360–63.

De Quervain, et al. (2004). The neural basis of altruistic punishment. *Science* 305(5688), 1254–58.

Singer, T., et al. (2006). Empathic neural responses are modulated by the perceived fairness of others. *Nature* 439(7075), 466–69.

Rockenbach, B., & Milinski, M. (2006). The efficient interaction of indirect reciprocity and costly punishment. *Nature* 444, 718–23.

Dabbs, J. M., & Dabbs, M. G. (2000). *Heroes, Rogues, and Lovers: Testosterone and Behavior.* New York: McGraw-Hill; Mehta, H., & Josephs, R. A. (2006). Testosterone change after losing predicts the decision to compete again. *Hormones and Behavior* 50(5), 684–92.

Johnson, R. T., & Breedlove, S. M. (2010). Human trust: testosterone raises suspicion. *Proceedings of the National Academy of Sciences* 107(25), 11149–50.

Lehrer, J. (May 18, 2011). How power corrupts. *The Frontal Cortex,* Wired Blogs, www.wired.com/wiredscience/2011/05/how-power-corrupts.

Burnham, T., McCabe, K., & Smith, V. L. (2000). Friend-or-foe intentionality priming in an extensive form trust game. *Journal of Economic Behavior & Organization* 43(1), 57–73.

Klucharev, V., et al. (2009). Reinforcement learning signal predicts social conformity. *Neuron* 61(1), 140–51.

Eisenberger, N. I., Lieberman, M. D., & Williams, K. D. (2003). Does rejection hurt? an fMRI study of social exclusion. *Science* 302(5643), 290–92.

Wagner, J. D., Flinn, M. V., & England, B. G. (2002). Hormonal response to competition among male coalitions. *Evolution and Human Behavior* 23(6), 437–42.

## CHAPTER 5: The Disconnected

Zak, P. J. (2005). Trust: a temporary human attachment facilitated by oxytocin. *Behavioral and Brain Sciences* 28(3), 368–69.

Harlow, H. F., & Zimmermann, R. R. (1959). Affectional responses in the infant monkey. *Science* 130, 421–32.

Wismer Fries, A. B., et al. (2005). Early experience in humans is associated with changes in neuropeptides critical for regulating social behavior. *Proceedings of the National Academy of Sciences* 102, 17237–40.

Autism—repetitive behaviors like rocking and flapping (2007). Retrieved from www.youtube.com/watch?v=f15JexiQt4U.

Sally, D., & Hill, E. (2006). The development of interpersonal strategy: autism, theory-of-mind, cooperation and fairness. *Journal of Economic Psychology* 27(1), 73–97.

Hamilton, John (August 23, 2010). Autism gives woman an "alien view" of social brains. *All Things Considered,* National Public Radio, www.wbur .org/npr/129379866/autism-gives-woman-an-alien-view-of-social-brains.

Hoge, E. A., et al. (2008). Oxytocin levels in social anxiety disorder. *CNS Neuroscience & Therapeutics* 14(3), 165–70, and additional unpublished data.

Uvnäs-Moberg, K., et al. (1999). Oxytocin as a possible mediator of S.S.R.I.-induced antidepressant effects. *Psychopharmacology* 142(1), 95–101.

Adolphs, R., Tranel, D., & Damasio, A. R. (1998). The human amygdala in social judgment. *Nature* 393, 470–74.

Greene, J. D., et al. (September 14, 2001). An fMRI investigation of emotional engagement in moral judgment. *Science* 293, 2105–08.

Lee, Henry K. Hans Reiser case: Nov. 19, 2008. *Local News Blog, San Francisco Chronicle* Online Edition, retrieved December 17, 2009, www.sfgate.com/ cgi-bin/blogs/localnews/detail?entry_id=32797.

Salter, A. (2004). *Predators: Pedophiles, Rapists, and Other Sex Offenders.* New York: Basic Books.

## CHAPTER 6: Where Sex Touches Religion

Harris, S. (2005). *The End of Faith.* New York: W. W. Norton.

Dawkins, R. (2008). *The God Delusion.* Boston: Mariner Books.

Dumont, G. J., et al. (2009). Increased oxytocin concentrations and prosocial feelings in humans after ecstasy (3,4-methylenedioxymethamphetamine) administration. *Social Neuroscience* 4(4), 359–66.

Wade, N. (2009). *The Faith Instinct: How Religion Evolved and Why It Endures.* New York: Penguin Press HC.

## CHAPTER 7: Moral Markets

Zak, P. J., editor. (2008). *Moral Markets: The Critical Role of Values in the Economy.* Princeton, NJ: Princeton University Press.

Zak, P. J. (2011). Moral markets. *Journal of Economic Behavior & Organization* 77(2), 212–33.

Hill, K. R., et al. (2011). Co-residence patterns in hunter-gatherer societies show unique human social structure. *Science* 331(6022), 1286–89.

Henrich, J., et al. (2005). "Economic man" in cross-cultural perspective: ethnography and experiments from 15 small-scale societies. *Behavioral and Brain Sciences* 28, 795–855.

Henrich, J., et al. (2010). Markets, religion, community size, and the evolution of fairness and punishment. *Science* 327, 1480–84.

Greif, A. (2006). *Institutions and the Path to the Modern Economy: Lessons from Medieval Trade.* New York: Cambridge University Press.

Osborne, D., & Gaebler, T. (1993). *Reinventing Government: How the Entrepreneurial Spirit Is Transforming the Public Sector.* New York: Plume.

Ott, J. (2005). Level and inequality of happiness in nations: does greater happiness of a greater number imply greater inequality in happiness? *Journal of Happiness Studies* 6(4), 397–420; Wilkinson, W. (April 11, 2007). In pursuit of happiness research: is it reliable? what does it imply for policy? *Cato Institute Policy Analysis Series* 590.

## CHAPTER 8: A Long and Happy Life

Montgomery, C. (2012). *Happy City.* New York: Farrar, Straus and Giroux.

Barrionuevo, A. (October 11, 2010). In rough slum, Brazil's police try soft touch. *New York Times.*

Haederle, M. (August 9, 2010). The best fiscal stimulus: trust. *Miller-McCune.*

Zak, P. J., & Knack, S. (2001). Trust and growth. *The Economic Journal* 111, 295–321.

Gelfand, M. J., et al. (2011). Differences between tight and loose cultures: a 33-nation study. *Science* 332(6033), 1100–104.

Knack, S., & Zak, P. J. (2002). Building trust: public policy, interpersonal trust, and economic development. *Supreme Court Economic Review* 10, 91–107.

Stout, H. (May 2, 2010). Antisocial networking? *New York Times.*

Penenberg, A. (July–August 2010). Dr. Love. *Fast Company*, 801–08.

DeParle, J. (October 9, 2010). Arizona is a haven for refugees. *New York Times.*

Greider, William. (1997). *One World, Ready or Not: The Manic Logic of Global Capitalism.* New York: Simon & Schuster.

Knack, S., & Zak, P. J. (2002). Building trust: public policy, interpersonal trust, and economic development. *Supreme Court Economic Review* 10, 91–107.

Kristof, N. D. (January 7, 2010). The happiest people. *New York Times.*

Zak, P. J., & Fakhar, A. (2006). Neuroactive hormones and interpersonal trust: international evidence. *Economics & Human Biology* 4, 412–29.

Brooks, Arthur C. (2008). *Gross National Happiness: Why Happiness Matters for America—and How We Can Get More of It.* New York: Basic Books.

Seligman, Martin E. P. (2004). *Authentic Happiness: Using the New Positive Psychology to Realize Your Potential for Lasting Fulfillment.* New York: Free Press.

Lay abstract: Merlin, R., Grosberg, D., & Zak, P. J. Oxytocin and happiness. Center for Neuroeconomics Studies, Claremont Graduate University.

# Acknowledgments

Innumerable generous people made this book, and the research behind it, possible. First among these is my wife, Lori, who spent many days without me as I traveled around the world doing research. She consistently encouraged me to continue my mission even though it meant hardships for her. My daughters, Alex and Elke, were often impatient for me to come home but persevered in my absence and were inevitably waiting for me at the door as I walked in, and tackled me. My parents, Donald and Dorothy Zak, gave me the gift of curiosity that made my journey possible, and provided me the love to survive the hardships I faced.

The incomparable William Patrick was my writing partner, critic, motivator, and now friend. This book would not be one-tenth as good without him. Absolutely essential to this endeavor were my brilliant agent Ms. Linda Loewenthal and my amazing lawyer Jeff Silberman, who put Bill and me together and provided sage advice throughout to keep the project moving forward. My editor at Dutton, Stephen Morrow, gave me the freedom and encouragement to write an unlikely science story about finding a new part of human nature and was enthusiastic about this book from our first meeting. Stephen, and Dutton president Brian Tart, never wavered in their belief in me and this project and have been fabulous in every aspect of the writing and publishing.

Many generous people and institutions provided funding for the research I have done. These include Dr. Jack Templeton, Dr. Barnaby Marsh, Dr. Kimon Sargeant, Dr. Paul Wason, and Mr. Chris Stawski of the John Templeton Foundation; Dr. Margaret Gruter and Ms. Monika Gruter Cheney of the Gruter Institute for Law and Behavioral Research; Mr. Gordon Getty of the Ann and Gordon Getty Foundation; Mrs. Victoria Seaver Dean of the Seaver Institute; Dr. Lis Nielsen of the National Institute on Aging; Mr. Gerry Ohrstrom, Mr. Skip Stein, and five presidents at Claremont Graduate University who directly enabled my work, Dr. Steadman Upham, Mr. William Everhart, Dr. Robert Klitgaard, Dr. Joseph Hough Jr., and Dr. Deborah Freund.

Those who took the most risks by working with me and did the lion's share of the research were my intrepid collaborators, Dr. Robert Kurzban, Dr. William Matzner, Dr. Stephen Knack, Dr. Jorge Barraza, Dr. Karla Morgan, Dr. Jang-Woo Park, Dr. Moana Vercoe, Dr. Vera Morhenn, Ms. Laura Beavin, Dr. Ahlam Fakhar, Ms. Beth Terris, Ms. Veronika Alexander, Dr. Sheila Ahmadi, Dr. Ronald Swerdloff, Dr. Walter Johnson, Dr. Cameron Johnson, Dr. Markus Heinrichs, Dr. Michael Kosfeld, Dr. Ernst Fehr, Dr. Urs Fischbacher, Dr. Bill Casebeer, Dr. Jeff Schloss, Dr. Michael McCullough, and Dr. Elizabeth Hoge.

Valued advisers who guided and often participated in my unusual research ideas include Dr. Yannis Venieris, the late Dr. Jack Hirshleifer, Dr. C. Sue Carter, Dr. Cort Pedersen, Dr. David

Levine, Ms. Estela Hopenhayn, Dr. Herb Gintis, Mr. Edward Tama, Ms. Linda Geddes, Mr. Nic Fleming, Dr. Helen Fisher, Dr. Michael McGuire, Dr. Lionel Tiger, Ms. Mary Jaras, Mr. Andrew Mayne, Lieutenant Colonel William Fitch, Professor Adam Penenberg, Dr. Michael Shermer, Dr. Matt Ridley, Mr. Kenshi Fukuhara, Mr. Itay Heled, Mr. Karl Jason, Ms. Stephanie Castagnier, and Professor Oliver Goodenough. They never said "impossible," and through their gentle wisdom they improved everything I thought to do.

Finally, there are many dear friends and colleagues who suffered for four years by reading and hearing endlessly about this book and gave me the extraordinary gift of their time, energy, and expertise. They have improved the ideas here and sharpened my thinking substantially. A short list includes Dr. Cameron Johnson, Dr. Vance Johnson, Ms. Joana Johnson, Dr. Walter Johnson, Dr. Sana Quijada, Dr. Earl Quijada, Mr. Paul Wheeler, Justice Thomas Hollenhorst, Mr. Tim Brayton, Mrs. Luzma Brayton, Dr. Thomas Borcherding, Dr. Thomas Willett, Dr. Arthur Denzau, Dr. Joshua Tasoff, Dr. Cyril Morong, Dr. Jeff Schloss, Dr. Paul Ingmundson, Dr. Michael Uhlmann, Dr. Jean Schroedel, Dr. Jacek Kugler, Dr. Gerald Winslow, Dr. Brian Bull, Dr. Carla Gober, Mr. Bruno Giussani, and Mr. Chris Anderson.

All those listed here, and many, many others, shared their love with me. I am grateful beyond measure.

# Index

# INDEX

Paul J. Zak has a Ph.D. in Economics from the University of Pennsylvania, and postgraduate training in Neuroscience from Harvard University. He is now Professor of Economics, Psychology and Management and Claremont, and Clinical Professor of Neurology and Loma Linda University Medical Center in California.

The son of a nun, Zak has a lifelong fascination with what motivates humanity and the biology of our decision-making. In 2002 he founded the first dedicated neuroeconomics laboratory in the world. He is the editor of *The Moral Markets: The Critical Role of Values in the Economy*, and his blog, 'The Moral Molecule', is a regular feature at PyschologyToday.com.